# PRIMARY SCIENCE READERS' THEATRE

*by Sharon Solomon*

*Illustrated by Greg Lawhun*

© 2002 Pieces of Learning

www.piecesoflearning.com
CLC0265
ISBN 978-1-931334-020
Printed by Total Printing Systems
Newton IL
5/2014

All rights reserved. In our effort to produce high quality educational products we offer portions of this book as "reproducible." Permission is granted, therefore, to the buyer - one teacher - to reproduce student activity pages in LIMITED quantities for students in the buyer's classroom only. The right to reproduce is not extended to other teachers, entire schools, or to school systems.
No other part of this publication may be reproduced in whole or part. The whole publication may not be stored in a retrieval system, or transmitted in any form or by any means, electronic, mechanical, photocopying, recording, or otherwise without written permission of the publisher.
For any other use contact Pieces of Learning at 1-800-729-5137.
For a complete catalog of products contact Pieces of Learning
or visit our Web Site at www.piecesoflearning.com

# TABLE OF CONTENTS

My Science Picture Dictionary . . . . . . . . . . . . . . . . . 6

Alfie Astronaut Visits the Moon . . . . . . . . . . . . . . . . 9
  . . A play to teach about the earth and the moon

The Case of the Spreading Germs . . . . . . . . . . . . . . 21
  . . A play to teach how to prevent germs from spreading

Dinosoccer . . . . . . . . . . . . . . . . . . . . . . . . . . . . . . 28
  . . A play to teach about plant-eating and meat-eating dinosaurs

How Flossy Flamingo Lost Her Habitat . . . . . . . . . . . 36
  . . A play to teach about living things and their habitats

How Rachel Rabbit Fixed the Food Chain . . . . . . . . . 46
  . . A play to teach about the food chains of animals

The Particles Visit Energyland . . . . . . . . . . . . . . . . . 54
  . . A play to teach how energy moves through matter

Stop! You're Interrupting My Life Cycle . . . . . . . . . . 63
  . . A play to teach the life cycles of plants and animals

The Sunbeams Take a Trip . . . . . . . . . . . . . . . . . . . 73
  . . A play to teach about light and color

Wally the Water Waster . . . . . . . . . . . . . . . . . . . . . 83
  . . A play to teach the sources of water and the importance
     of water conservation

What Did You Eat For Breakfast? . . . . . . . . . . . . . . 92
  . . A play to teach about healthy eating and nutrition

Why Matter Matters . . . . . . . . . . . . . . . . . . . . . . . 103
  . . A play to teach the forms of matter

# FROM THE AUTHOR

Primary Science Readers' Theatre was designed to be a unique way to integrate science curriculum with reading, writing, music, and art. As a Reading Specialist, I realized how important it was to bring science topics into the realm of literature. The characters are humorous and fictitious, but the vocabulary and science ideas are real.

The science topics correlate to these National Science Content Standards grades K-4.

All students should develop:

- abilities necessary to do scientific inquiry

- an understanding about scientific inquiry

- an understanding of properties of objects and materials

- an understanding of light, heat, electricity, and magnetism

- an understanding of life cycles of organisms

- an understanding of organisms and environments

- an understanding of properties of earth materials

- an understanding of objects in the sky

- an understanding of changes in earth and sky

- an understanding of personal health

- an understanding of types of resources

- an understanding of changes in environments

All prereading, purposes for reading, and post-reading activities are correlated with Bloom's Taxonomy. I hope you and your students enjoy the plays and the activities.

# TO THE TEACHER

You are about to engage your students in Readers' Theatre.

Readers' Theatre is a way to involve all students in a choral reading manner. Each play has enough parts for all of the children in your class. Suggestions are made for costumes, props, and scenery, but none are necessary for a successful presentation. The audience can use their imaginations as the students read their parts.

Before each play, there are prereading activities. You may choose to use the Science Picture Dictionary Forms. Duplicate enough pages so that words can be entered in alphabetical order. Students will write the word and its definition. They can then draw a picture in the box directly across from the entry. Vocabulary can be done orally without the picture dictionary if you so choose.

Each play has prereading activities and a purpose for reading which correlate with Bloom's Taxonomy. After reading the play, there are three post-reading activities to choose from. Each play has one activity sheet that accompanies a post-reading activity.

You will find the plays very easy to follow. The activities do not require materials other than writing and drawing paper, crayons and pencils. Students will be learning science vocabulary and concepts through reading, writing, music and art.

The songs included in each script are reinforcement of concepts in yet another learning style. Let the fun begin!

Primary Science Readers' Theatre

# MY SCIENCE PICTURE DICTIONARY

1. _____

_____

_____

2. _____

_____

_____

3. _____

_____

_____

4. _____

_____

_____

5. _____

_____

_____

6. _____

_____

_____

© 2002 Pieces of Learning

Primary Science Readers' Theatre

# MY SCIENCE PICTURE DICTIONARY

7. _____

_____

_____

8. _____

_____

_____

9. _____

_____

_____

10. _____

_____

_____

11. _____

_____

_____

12. _____

_____

_____

© 2002 Pieces of Learning

# MY SCIENCE PICTURE DICTIONARY

13. _____

_____

_____

14. _____

_____

_____

15. _____

_____

_____

16. _____

_____

_____

17. _____

_____

_____

18. _____

_____

_____

# ACTIVITIES FOR "ALFIE ASTRONAUT VISITS THE MOON"

## Prereading Activities

Students can enter their vocabulary into their Science Picture Dictionaries or the teacher can orally review the words with the class before reading the play.

revolution    rotation    axis    gravity    orbit

atmosphere    Polaris    tether    phases    meteorite

Explain why an astronaut would need a spacesuit on the moon. Write at least two reasons.

## Purpose For Reading

Predict what it would be like for you to visit the moon. Discuss with your neighbor what you might see and do on the moon.

## Post-reading Activities

Draw and label what Alfie saw on the moon. At the bottom of the picture, write a few sentences telling what he saw and did on the moon.

Using Activity Sheet #1, compare and contrast the Earth and the Moon.

Research a space flight or an astronaut. Use books, encyclopedias, or the internet. NASA has a home page with lots of information about its space missions. Use Activity Sheet #2 to help you.

Primary Science Readers' Theatre

# ACTIVITY SHEET #1
# "ALFIE ASTRONAUT VISITS THE MOON"

Name _____

Compare and contrast the Earth and the Moon.

How they are the same:

_____
_____
_____

How they are different:

_____
_____
_____

I would rather live on _____ because _____

_____
_____
_____

# ACTIVITY SHEET #2
# "ALFIE ASTRONAUT VISITS THE MOON"

Name _____

Circle your answer.

I plan to research      an astronaut      or      a space mission.

The name of my astronaut or space mission is _____
_____

List five exciting facts about your astronaut or space mission.

1. _____
   _____

2. _____
   _____

3. _____
   _____

4. _____
   _____

5. _____
   _____

Rewrite this information into a report.
Design a cover with a diagram or picture that goes with your report.

Primary Science Readers' Theatre

# "ALFIE ASTRONAUT VISITS THE MOON"
A play to teach about the earth and the moon

Characters

Commander Revolution     Commander Rotation     Alfie Anderson
Narrator     Miss Universe (teacher)     Class

Costumes   three helmets and backpacks, class wears t-shirts, Commanders Revolution and Rotation have their names on their shirts.

Props  a globe, a paper sun, a chart with the planets

Vocabulary  revolution, rotation, axis, gravity, orbit, meteorite, atmosphere, Polaris, tether, phases

> SCENE 1   North Star Elementary School, Polaris, U.S.A.

**Narrator**: Miss Universe is returning science tests to her class. The tests were about the Moon and the Earth. As she hands back the test to Alfie Anderson, Miss Universe makes an announcement.

**Miss U**:  Class, remember I told you there would be a surprise today? Well, the winner of the free trip to the Moon is Alfie Anderson. Alfie got a low grade on our science test about the Moon. Of all the students in our class, who needs to learn about the Moon the most?

**Class**:  Alfie Anderson!

**Miss U**:  That's right!

**Narrator**:  Alfie is daydreaming, as usual. Papers are stuffed in his desk and are spilling onto the floor.

**Alfie**: What? What's that you say?

**Miss U**:  Class, tell Alfie where he's going.

**Class**: To the Moon.

**Miss U**: And to escort you to the Moon, I am sending along Commander Revolution and Commander Rotation.

**Class**: Hooray!

**Alfie**: Rotation? Revolution? Huh?

**Narrator**: The two commanders enter the room.

**C. Revolution**: Alfie, come along. The rocket is waiting out on the playground.

**C. Rotation**: Put on your spacesuit and helmet and we'll get ready for takeoff.

**Alfie**: This spacesuit sure is heavy!

**C. Revolution**: Well, Alfie, our suits weigh about 100 pounds. Yours must weigh about 80 pounds.

**Alfie**: Am I dreaming or going crazy?

**C. Rotation**: Alfie, weren't you listening when Miss Universe said you won the free trip to the Moon?

**C. Revolution**: Let's go, buddy.

**Class**: Have a good trip!

**Narrator**: Alfie, Commander Revolution and Commander Rotation climb into the spaceship.

**Alfie**: Can I call Mom and Dad to let them know I won't be home for supper?

**Commanders**: Miss Universe took care of that. Relax, Alfie. You're in for the ride of your life!

**Alfie**: I'm getting you two commanders mixed up. How can I tell you apart?

**Narrator**: Commander Revolution points to the sun on the chart. Commander Revolution traces the Earth's orbit around the sun.

**C. Revolution**: Well, I'm Commander Revolution. The Earth revolves around the sun. It travels in an orbit that takes 365 days to complete. That is one revolution.

**Alfie**: So, all this time I've been on Earth we've been orbiting the sun?

**Commanders**: Yes, Alfie, but you weren't listening to Miss Universe.

**Alfie**: What is rotation?

**C. Rotation**: (Commander Rotation spins around once.) While the Earth is in orbit around the sun, it is also turning on its axis.

**Alfie**: Taxes? Faxes?

**Commanders**: Axis.

**Narrator**: Commanders Revolution and Rotation point to the globe and turn it.

**C. Rotation**: There is an imaginary line called an axis.

**Alfie**: Axis?

**Commanders**: Right!

**C. Rotation**: The Earth turns on its axis every 24 hours.

**Alfie**: So, when North Star Elementary School is facing the sun it's daytime and when it's turned away from the sun, it's nighttime?

**Commanders**: You got it!

**Miss U**: Class, look out the window. They have to travel 384,000 kilometers to get to the Moon.

**Class**: That's far!

**Alfie**: The Earth is looking smaller and smaller.

**Narrator**: Alfie and Commanders Revolution and Rotation are on their way to the Moon.

SCENE 2   On the Moon

**Narrator**: The spacecraft lands gently. Alfie and the commanders put on their backpacks and hook up the oxygen and water tubes.

**C. Rotation**: Be careful climbing down the ladder, Alfie.

**C. Revolution**: There is very little gravity on the Moon, so take small steps. If you don't, you'll get too far away from us.

**Narrator**: Alfie climbs down the ladder and begins to jump. He goes way up into the air.

**Alfie**: Help! Help! Get me down!

**Narrator**: Commander Rotation grabs Alfie's foot and pulls him back down.

**C. Rotation**: On the Moon, you can jump six times higher than on the Earth. You don't weigh as much because there is much less gravity.

**Alfie**: Cavity?

**Commanders**: Gravity. Don't you ever listen to Miss Universe?

**C. Revolution**: Alfie, gravity is the force that causes objects to move toward each other. That's what kept your feet on the ground on Earth.

**C. Rotation**: Watch where you're walking. It's very dark and you could trip. Watch out for craters.

**Alfie**: Gaters?

**Commanders**: Craters.

**C. Revolution**: Craters are pits or holes in the ground.

**C. Rotation**: Craters are shaped like bowls. Meteorites hit the surface of the Moon and make holes where they crash.

**Alfie**: Meteor - nites? Meteor-lites?

**Commanders**: Meteorites.

**C. Revolution**: Meteorites are chunks of rock or metal that fall from space.

**Alfie**: I hope one isn't headed this way.

**C. Rotation**: Meteorites can be small or as big as 1,000 kilometers.

**Alfie**: I can see the Earth and the stars! It looks so dark out there!

**C. Revolution**: That's because the Moon has no atmosphere.

**Alfie**: Atmosphere?

**C. Rotation**: There are no gases surrounding the Moon.

**C. Revolution**: There is also no water, so nothing could really live here without air or water.

**Alfie**: It's a good thing we're wearing our spacesuits. At least we can breathe.

**C. Rotation**: Yes, you have an oxygen tank in your backpack.

**C. Revolution**: And a tube inside your helmet for drinking water.

**Alfie**: It's a good thing we're tied down, too. I wouldn't want to fly away.

**C. Rotation**: The rope is called a tether.

**Alfie**: Weather? Heather?

**Commanders**: Tether.

**Alfie**: I know this spacesuit was heavy on Earth. Why doesn't it feel heavy on the Moon?

**C. Rotation**: The Earth has more mass than the Moon, so it has more gravity.

**Alfie**: Good! I won't ever have to go on a diet.

**Narrator**: Alfie looks puzzled and asks a question.

**Alfie**: I see that the Earth looks like a ball from here, but why doesn't the Moon always look like a ball from Earth?

**Commanders**: Phases.

**Alfie**: Phrases? Vases?

**Commanders**: Phases.

**C. Rotation**: The Moon also has an orbit. It rotates around the Earth.

**C. Revolution**: As it rotates, we see different amounts of the daylit half of the Moon.

**Alfie**: The Moon rotates, too! Nothing stands still. But it doesn't feel like we are moving.

**C. Rotation**: Each phase of the Moon has a name. There is a new moon, a first quarter, a full moon and a last quarter.

**C. Revolution**: The New Moon occurs when the side of the Moon facing the Earth receives no sunlight at all.

**Alfie**: The Moon has less gravity and its light is reflected from the sun. Does the Moon have any power of its own?

**C. Rotation**: Well, as it rotates around the Earth, its gravity controls the tides of the oceans.

**C. Revolution**: The side of the Earth closest to the Moon will have high tide and at the same time, a second high tide occurs on the side of the Earth farthest from the Moon.

**C. Rotation**: Every 24 hours there are 2 high tides and 2 low tides for each ocean.

**Alfie:** Far out!

**Commanders**: That's for sure.

**Alfie**: I'm getting homesick - I mean Earthsick. Can we go back now? I promise to pay attention in class and to study for my science tests.

**Commanders**: Away we go!

**Narrator**: Alfie and Commanders Revolution and Rotation climb back into the spaceship and ride the 384,000 kilometers to Earth.

SCENE 3   North Star Elementary School, Polaris, U.S.A.

**Narrator**: The spaceship lands on the playground of North Star Elementary School. Miss Universe and the class are looking out the window as Alfie and the Commanders climb out of the spaceship.

**Alfie**: That was very exciting. Thank you. Can I get out of my spacesuit now? It's so heavy.

**Commanders**: Sure. Let's go tell the class all about your trip.

**Narrator**: Alfie and Commanders Revolution and Rotation enter the classroom.

**Class**: Hooray! Welcome back. How was your trip?

**C. Revolution**: Well, Alfie almost got away from us.

**Alfie**: There is so little gravity on the Moon. I had to take baby steps.

**Miss U**: Did you pay attention, Alfie?

**Alfie**: I certainly did, Miss Universe. It's not every day you get a chance to go to the Moon.

**Class**: What did you see?

**Alfie**: Well, there were craters. It was cold and dark on the side where we landed. We had oxygen and water in our spacesuits because there is none on the Moon.

**Miss U**: Do you understand revolution and rotation?

**Narrator**: Alfie puts a paper sun on the floor. He walks slowly around the sun.

**Alfie**: The Earth revolves around the sun. It travels in its orbit. It takes a year to go one revolution around the sun.

**Class**: What about rotation, Alfie?

**Alfie**: Rotation is the Earth turning on its axis. (Alfie spins around once.) This happens every 24 hours. That's why we have day and night.

**Miss U**: Well, Alfie, it looks like you learned a lot.

**Commanders**: Alfie had a little trouble at first, but he adjusted nicely to his new environment.

**Class and Miss U**: Congratulations, Alfie.

**Commanders**: Before we go, we have a song for you. Let's sing along together.

*(To the tune of "Yankee Doodle")*

Alfie Anderson went to the Moon,
Riding on a spaceship.
Stuck some oxygen on his back
And went upon a space trip.

Alfie Astronaut, keep it up, Alfie Anderson.
Pay attention to Miss U
And get your schoolwork done.

**Class and Miss U**: So long, Commanders Revolution and Rotation.

**Narrator**: As the two commanders leave, Alfie thinks out loud.

**Alfie**: Mom and Dad will never believe this. They already think I'm spaced out!

## THE END

# ACTIVITIES FOR "THE CASE OF THE SPREADING GERMS"

## Prereading Activities

Students can enter their vocabulary into their Science Picture Dictionaries or the teacher can orally review the words with the class before reading the play.

bacteria   germs   microscope   prevention

Have students tell a partner or share with the class what they do to stay healthy.

## Purpose For Reading

Find out what the students at Smiley Elementary School learn about staying healthy from Captain Healthy and Captain Prevention.

## Post-reading Activities

Make a list of ways to prevent germs from spreading. Give the list to each student who will then tally how many times in one week they practiced these rules. Make the students' tallies into a bar graph or a line graph.

Make a plan about how they will change one room in their house to make it healthier. They can diagram and label the changes on the activity sheet for *The Case of the Spreading Germs*.

Musically inclined students may wish to compose another stanza to the song at the end of the play.

## ACTIVITY SHEET
## "THE CASE OF THE SPREADING GERMS"

Draw a diagram of a room in your house that shows how you will change it to make it healthier. Label all of the changes. Try to make at least three changes to the room to prevent germs from spreading.




My _____ room is healthier now because _____

_____

_____.

Primary Science Readers' Theatre

# "THE CASE OF THE SPREADING GERMS"
A play to teach how to prevent germs from spreading

Characters
Captain Healthy      Captain Prevention      Miss Mucus
Narrator             Emily                   Krista
Anna                 Blake                   Mike           Class

Costumes  Students wear school clothes. Captains Healthy and Prevention wear shirts with capital "H" or capital "P" and capes.

Props  scissors

Scenery  Classroom, playground with swings

Vocabulary  bacteria, germs, microscope, prevention

    SCENE ONE     *Smiley Elementary School, Monday morning*

**Miss Mucus**: Well, class, everyone is here today except Ted. Ted's dad called to say he has a fever and a sore throat.

**Krista**: I had that on Friday, but Mom sent me to school anyway. She said I'd feel better.

**Mike**: I remember you were coughing on Friday.

**Emily**: Are you feeling better?

**Krista**: Yes, but I went to the doctor after school. She said I had bacteria that made me sick. I'm taking special medicine to get better.

**Miss Mucus**: You sat next to Ted on Friday. I wonder if the bacteria that made you sick is why Ted has a fever and a sore throat.

**Krista**: Oh, I don't think so. How could bacteria get from my throat into Ted's throat?

© 2002 Pieces of Learning

**Narrator**: Just as Miss Mucus was about to explain, there was a flash of lightning and a puff of smoke. When the air cleared, two strange creatures in capes flew into the room.

**Class**: Who are you?

**Captain Healthy**: I'm Captain Healthy.

**Captain Prevention**: And I'm Captain Prevention.

**Class**: W-w-why are you h-h-here?

**Both Captains**: We're here to help the kids of Smiley School stay healthy.

**Captain Healthy**: Too bad we were too late for Krista and Ted.

**Captain Prevention**: Let's go outside and we'll show you how bacteria and germs are spread.

*SCENE II   Playground, Smiley Elementary School*

**Narrator**: Miss Mucus and her class follow Captain Healthy and Captain Prevention outside to the playground.

**Krista**: Why are we going outside? I don't see anything special on the swings.

**Class**: What are we looking for? What do germs look like?

**Captain Prevention**: That's just the problem. You can't see germs or bacteria unless you look at them under the microscope.

**Class**: What do they look like? Where can we get some germs to look at?

**Captain Healthy**: See the swings over there? Well, Mr. Saliva's class was just out here playing.

**Captain Prevention**: I saw Jerry sneezing into his hands. Then he went on the swings.

**Mike**: So, I've done that lots of times.

**Class**: Me, too.

**Captain Healthy**: We'll rub this clean cloth on the chains of the swings.

**Captain Prevention**: Now, let's go back inside and look at the cloth under the microscope.

SCENE III       Miss Mucus's class

**Narrator**: Back in Miss Mucus's class, Captain Healthy and Captain Prevention put the cloth under the microscope. The class takes turns looking at the brightly colored squiggly and round shapes.

**Class**: These are weird!!

**Blake**: Billy, come look at these orange circles!

**Captains Healthy And Prevention**: You have just seen what germs look like.

**Class**: Wow!

**Emily**: Anna, I saw you sneezing when you came in this morning.

**Anna**: So what? I didn't sneeze on anyone. I covered my nose and mouth just like Miss Mucus taught us to.

**Captains Healthy And Prevention**: Yes, but you used those scissors and then Blake used them.

**Class**: EEWWWW!

**Mike**: Let's put the scissors under the microscope!

**Krista**: Wow, I see red swirls and green stuff.

**Class**: Yuck!

**Captain Prevention**: What could you have done, Anna, to prevent your germs from spreading?

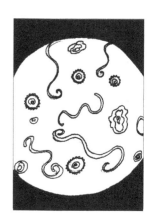

**Anna**: I covered my nose and mouth with my hands.

**Blake**: Yes, but your germs went on the scissors and I used them.

**Class**: That's disgusting!!

**Captain Prevention**: You should have washed your hands with soap and water.

**Captain Healthy**: You could have sneezed into a tissue.

**Blake**: Well, I'm going to wash my hands with soap and water right now to get rid of the germs.

**Emily**: Krista, I think next time you have a fever and sore throat you should stay home. Poor Ted. I wonder how he's doing.

**Krista**: Now I see why Miss Mucus thinks I might have spread my bacteria to Ted. He borrowed my pencil and I had put it in my mouth just before he used it.

**Captain Healthy**: To stay healthy, don't put things in your mouth.

**Captains Healthy And Prevention**: We want to leave with this little song. After we sing it, please join in and we'll sing it together.

ALL: *(To the tune of "A, B, C, D, E, F, G," etc.)*

Don't sneeze germs into the air,
Use a tissue, show you care.
Wash your hands so germs won't spread,
Or you will be sick like Ted.
To prevent germs from your sneeze,
Be more careful, won't you please?

**Narrator**: With a puff and a flash, Captain Healthy and Captain Prevention fly away.

**Class**: Where did they go?

**Narrator**: The children and Miss Mucus run to look out the window.

**Miss Mucus**: It looks like they are flying near Happy Elementary School. I'll bet they have a lot of germs over there.

**THE END**

# ACTIVITIES FOR "DINOSOCCER"

## **Prereading Activities**

Students can enter their vocabulary into their Science Picture Dictionaries or the teacher can orally review the words with the class before reading the play.

defense   extinct   forwards   fossils   midfielders   remains   imprints

Ask students to draw a soccer field and label where the players would be on the field. Discuss the rules of soccer and make a class list of the rules of the game.

## **Purpose For Reading**

Think about the rules for playing soccer. Imagine dinosaurs playing soccer and predict who you think will win - the plant-eaters' team or the meat-eaters' team. Give at least one reason for your prediction.

## **Post-reading Activities**

Research a dinosaur by assigning a different one to each student. Students can use books or search on the computer to find information about the size, body, habitat and food of their dinosaurs. (See *Dinosaur Research Activity Sheet*.)

Make a class line or bar graph bulletin board. Each student will supply data on the height and weight of their dinosaur to be plotted onto the graph. Ask students to conclude which dinosaur was the tallest, heaviest, smallest, and lightest in weight.

Play "Question the Expert." Each student will show a picture of their assigned dinosaur and tell the class five interesting facts about it. The student will sit in the "expert chair" and the class can ask questions about that dinosaur.

Primary Science Readers' Theatre

# Activity Sheet
# Dinosaur Research

Name_____

Name of dinosaur _____

Height_____

Weight_____

Special body features_____
_____

Food_____

Habitat_____

Draw a picture of your dinosaur in its habitat.

29

© 2002 Pieces of Learning

Primary Science Readers' Theatre

# "DINOSOCCER"
A play to teach about plant-eating and meat-eating dinosaurs

Characters

Narrator   George   Michelle   Allosaurus (al o SAUR us)
Brachiosaurus (Brack e o SAUR us)   Diplodocus (di PLOD u cus)
Ornitholestes (or ni tho LESS tes)   Protoceratops (pro to SAIR o tops)
Triceratops (tri SAIR o tops)   Tyrannosaurus (ti ran o SAUR us)
Velociraptor (vel law si RAP tor)
Meat-eaters: Allosaurus, Ornitholestes, Tyrannosaurus, Velociraptor
Plant-eaters: Brachiosaurus, Diplodocus, Protoceratops, Triceratops

Vocabulary  defense, extinct, forwards, fossils, imprints, mid fielders, remains

## SCENE I

**Narrator**: One rainy day, George and his sister Michelle were in the family room looking for something to do.

**Michelle**: George, it's been raining all morning and our soccer games are canceled.

**George**: I wish I were playing soccer with my team right this minute.

**Michelle**: Let's play with your dinosaur models. We've been learning about dinosaurs at school. I can tell the plant-eaters from the meat-eaters.

**George**: I know, the ones with the long sharp teeth are the meat-eaters. The plant-eaters have flat teeth for grinding up the plants they eat.

**Michelle**: Let's set the dinosaurs up on the ping-pong table.

© 2002 Pieces of Learning

**George**: Michelle, it looks like a soccer field. Let's pretend that the dinosaurs are playing soccer!

**Michelle**: How about the meat-eaters against the plant-eaters? If biting is not allowed, it would be a fair game.

**Narrator**: George and Michelle chose the dinosaurs and placed the small, faster ones in front as forwards, the strong ones as mid fielders, and the larger ones in the back for defense.

**George**: I'm using dinosaur remains to make the goal and a piece of Mom's stocking for the net. The remains are just fake bones like the real ones scientists have found, only smaller.

**Michelle**: Are we ready to play?

**George**: I have a mini soccer ball. Remember, only the goalie can use his hands. No biting is allowed. May the best dinosaurs win!

**Narrator**: At the kickoff, there was a loud noise and the table began to shake. George and Michelle crouched in the corner and watched the dinosaurs come to life.

SCENE II    *The soccer game*

**Narrator**: George had set up the plant-eaters on one side and the meat-eaters on the other side. The plant-eaters' team was triceratops, the goalie, 3 protoceratops forwards, 3 diplodocus mid fielders, and 4 brachiosaurus defense players.

**Triceratops**: I guess George and Michelle picked me because of my mighty three horns and my skull like a giant shell.

**Protoceratops**: Listen, Triceratops, you're also a big guy. You're 30 feet long and weigh 11 tons. We Protoceratops will make great forwards because we're small and fast. We're only 6 feet long and weigh 400 pounds. We'll be able to score some goals.

**Diplodocus**: Protoceratops, your snout sticks out and will be good for heading the ball. We are 100 feet long and weigh 11 tons. We have VERY long necks and tails. We'll be able to pass the ball to you easily.

**Brachiosaurus**: We may be shorter than you a bit, but we weigh 89 tons. We're heavier than 12 elephants! We have VERY long necks and BIG feet. We'll be able to get the ball away from those meat-eaters. We can raise our heads over 40 feet high.

**Narrator**: The meat-eaters laughed as they listened to the plant-eaters.

**Meat-Eaters**: Ha! Ha! Ha! Listen to those plant-eaters! We'll win this soccer game and then eat them for dessert!

**Narrator**: The meat-eaters' team was allosaurus, the goalie, 3 velociraptor forwards, 3 ornitholestes mid fielders, and 4 tyrannosaurus defense players.

**Allosaurus**: I know I'll make a SUPER goalie because of my short powerful arms and my large claws. I'm only 36 feet long but I weigh 2 tons and can catch anything.

**Velociraptor**: We're only 6 feet long and weigh a little more than 200 pounds. We move VERY fast when we hunt, so we can run quickly to score goals. We'll make great forwards!

**Ornitholestes**: You Velociraptors have front and back claws, too. We're about 7 feet long but weigh just 35 pounds. We aren't heavy but we have BIG tails and BIG feet. So we can pass the soccer ball to you quickly.

**Tyrannosaurus**: We are 45 feet long and weigh 7 tons. Look at our 7-inch teeth. We could tear those plant-eaters apart before you can say "tyrannosaurus." Our huge tails will help us get the ball away from those plant-eaters. We'll be the best defense ever!

**Meat-eaters**: The plant-eaters kicked off. Brachiosaurus' big feet are enormous! Let's take a bite out of their necks.

**Narrator**: After lots of kicking and noise, the referee called a foul.

**Tyrannosaurus**: All right, all right. We know. No biting. We get it.

**Brachiosaurus**: We'll pass the ball to those Diplodocus dudes.

**Diplodocus**: Our tails can whack the ball over to our forwards, Protoceratops.

**Protoceratops**: Allosaurus isn't paying attention. We'll just kick the ball into the net. Hooray! Hooray! A goal!

**Meat-eaters**: It's not fair. Biting should be allowed!

**Plant-eaters**: When we're through, those meat-eaters will be fossils before you can say "Triceratops"!

**Tyrannosaurus**: Yippee! We just got the ball! We'll just use our HUGE tail to pass it to Ornitholestes. Hey, Orni! Watch out! Here it comes!

**Ornitholestes**: Wow! Here come those Diplodocus and Brachiosaurus dudes! We're so tiny. We better run fast. Our tail can pass the ball to the goalie. Pay attention, Velociraptors!

**Velociraptors**: We'll catch it with our back claws. We've got to get it past Triceratops, which won't be easy with its big horns and armor.

**Narrator**: Triceratops hit the ball with its horns. It bounced back and the Velociraptors quickly kicked the ball into the net. It was a tie score – one to one!

**All dinosaurs**: What's that noise and shaking? It's an earthquake! Run for your lives before we become extinct!


SCENE III    Back home


**George and Michelle**: What a mess!

**Michelle**: All that's left are a few dinosaur imprints and remains!

**George**: So, we don't know who won the soccer game.

**Michelle**: Guess we'll never know. George, I hope it doesn't rain next Saturday and we get to play soccer with our teams.

**George**: Me, too. But if it does rain, I thought we could set up some dinosaur models and play soccer again.

**Michelle**: Quit while you're ahead. Let's sing some songs until the rain stops. I've got one for you about dinosaurs.

*(To the tune of "Three Blind Mice")*

*Three Protoceratops,
three Protoceratops,
See how they run,
see how they run.*

*They all ran after the soccer ball
And scored a goal without a fall.
Three Protoceratops,
three Protoceratops.*

*Three Velociraptors,
three Velociraptors,
See how they run,
see how they run.*

*They all ran after the soccer ball
And scored because they were very small.
Three Velociraptors,
three Velociraptors.*

*Dinosaurs,
dinosaurs,
See how they run,
see how they run.*

*They all ran after the soccer ball
And now what's left isn't much at all.
Dinosaurs,
dinosaurs.*

**George**: Michelle, how about a nice quiet game of Hangman?

## The End

# ACTIVITIES FOR
# "HOW FLOSSY FLAMINGO LOST HER HABITAT"

### Prereading Activities

Students can enter their vocabulary into their Science Picture Dictionaries or the teacher can orally review the words with the class before reading the play.

    habitat    woodland    arctic    desert

Pretend you and your family are moving to an arctic habitat. List three changes you would have to make in your life to adjust to an arctic habitat. Write whether you think it would be better or worse than where you now live. Tell why you would feel this way.

### Purpose For Reading

Read to find out how Flossy Flamingo feels about going to a different habitat.

### Post-reading Activities

Draw a picture of Flossy in her water habitat. Label the plants and animals. (See *How Flossy Flamingo Lost Her Habitat* Activity Sheet).

Explain why flamingos need to live in shallow water. Give at least two reasons.

Would you rather live in a desert habitat or a woodland habitat? Describe three things you could do that you cannot do in your own habitat.

Primary Science Readers' Theatre

# ACTIVITY SHEET
# "HOW FLOSSY FLAMINGO LOST HER HABITAT"

Here is a picture of Flossy Flamingo. Draw her in her habitat and label the plants and animals you would find there.

Flossy Flamingo is happy to be back in her own habitat because

_____

_____

_____

© 2002 Pieces of Learning

Primary Science Readers' Theatre

# "HOW FLOSSY FLAMINGO LOST HER HABITAT"
A play to teach about living things and their habitats

Characters
Flossy Flamingo        Mother Flamingo        Father Flamingo
Freddy Flamingo        Narrators 1-5          Danny Deer
Deer Family            Paula Polar Bear       Polar Bear Family
Cathy Cactus           Carl Cactus

Backgrounds match the habitats   water, desert, arctic, and woodlands

Costumes   Flamingos - pink shirts; deer - brown shirts; polar bears - white shirts; cactus - green shirts

Vocabulary   habitat, woodland, arctic, desert

SCENE 1         A shallow lake in Florida

**Narrator 1**: One hot summer day, the Flamingo family was wading in the pond and eating lunch.

**Mother**: I just scooped up some yummy plants!

**Father**: These little fish taste SO good.

**Freddy**: Flossy, go scoop up your food over on the other side of the lake. This is MY side.

**Flossy**: It's my side. You go scoop your food over there.

**Freddy**: Will not!

**Flossy**: Will so!

**Freddy**: Will not!

**Flossy**: Will so!

**Mother**: Children! Stop fighting. There's plenty of food for all of us.

**Narrator 1**: Just then, Freddy pushed Flossy, and she fell face down into the lake. Flossy and Freddy began pushing and making lots of noise.

**Father**: Flossy, if this doesn't stop, there will be no dinner for you tonight. Leave your little brother alone!

**Flossy**: It's Freddy's fault. Just because I'm older doesn't mean it's my fault. I'm leaving, and I'm never coming back!

**Narrator 1**: Flossy Flamingo flew off, leaving her shocked family by the edge of the lake.

**Freddy**: Don't worry, Mom and Dad, she'll be back.

SCENE 2     *A woodland habitat in New Jersey*

**Narrator 2**: Flossy Flamingo was so upset she flew and flew until she got tired. When it got dark, she flew down and slept under a tree. When Flossy woke up, she did not know where she was.

**Flossy**: I'll just fly back home now. Everyone will miss me, and Freddy will behave himself.

**Narrator 2**: As Flossy Flamingo flew, she lost her way. She flew further and further away from Florida.

**Flossy**: I'm so tired and hungry. I've got to find a river or lake so I can look for food. Oh, I see something green down there. I'll just go see what it is.

**Narrator 2**: Flossy landed in a woodland habitat in New Jersey. Danny Deer and his family were eating flowers and plants, and Wanda Woodpecker was drilling the tree bark looking for food.

**Deer Family**: These daisies are delicious. Summer is such a nice time to live in the woods.

**Dan Deer**: Yes, we can eat all of the ferns and leaves in the forest.

**Deer**: What's that landing over there?

**Dan Deer**: It's bright pink and SO skinny!

**Deer**: Did you ever see legs like that?

**Dan Deer**: Its feet are webbed almost like a duck.

**Narrator 2**: Flossy is having trouble walking in the woods with her webbed feet.

**Flossy**: Help! Help! I'm falling! How is a beautiful wading bird like me supposed to walk on all this green stuff?

**Deer**: That green stuff is grass.

**Flossy**: I'm hungry. I need to find a river or a pond.

**Dan**: Bad luck! We do have a creek about a mile from here.

**Flossy**: Thanks! I'll fly there and eat my dinner.

**Deer**: Don't forget to write!

**Narrator 2**: When Flossy Flamingo got to the creek, it was only a few inches deep. Flossy didn't have enough water to scoop up her food.

**Flossy**: Yuck! All I'm getting is mud and rocks in my bill. The water isn't deep enough. I'll have to fly away from here. This woodland habitat is no place for a beautiful flamingo like me.

### SCENE 3  An arctic habitat

**Narrator 3**: Flossy Flamingo flew away from the woodland habitat. She thought she was on her way back to Florida, but she got really mixed up. Flossy Flamingo was flying to the North Pole!

**Flossy**: Br-r-r. It's so cold here. I'm tired and hungry. I'll just swoop down and look for food and shelter.

**Polar Bears**: What's that skinny pink thing that just flew over us?

**Paula**: Beats me! I've never seen anything like it. Let's make noises and jump up and down. Maybe the pink thing will come down here.

**Polar Bears**: Hello, pink flying thing. Come down here. We want to meet you.

**Flossy**: Oh, good. Maybe those chubby white guys can help me.

**Narrator 3**: So Flossy Flamingo landed on the ice in the cold arctic habitat.

**Flossy**: Hello, white things. Who are you?

**Polar Bears**: We're polar bears. Who are you?

**Flossy**: I'm Flossy Flamingo, and I'm a bird. I live near a lake in Florida. I like warm places, but I need to find a lake or a river so I can go scooping for my dinner. My bill is bent, and that's the way I eat.

**Paula**: We have plenty of water, but it's frozen solid, even in summer.

**Flossy**: Br-r-r. I'm freezing. I could use some long underwear. Got any?

**Polar Bears**: Ha, ha, ha!! We don't need long underwear, Flossy. We have thick fur to keep us warm.

**Paula**: Our white fur is good for hiding in the snow. We wait until the seals come out to hunt them for dinner.

**Narrator 3**: Flossy Flamingo tries to peck the ice with her bill.

**Flossy**: Ouch! This water is hard and cold. How can a beautiful bird like myself find dinner here?

**Paula**: You better fly south before you freeze to death. It's thirty degrees below zero here in our arctic habitat.

**Polar Bears**: Don't forget to write!

**Narrator 3**: So Flossy Flamingo left the arctic habitat. She was getting more tired and hungry, but she knew she could not live in the cold arctic temperatures.

SCENE 4    *A desert habitat*

**Narrator 4**: Flossy did fly south and west. Soon she saw large tan places with very few plants.

**Flossy**: Thank goodness it's getting warmer. I love hot places. I can't wait to see Mom and Dad, and even Freddy. I hope I can find a river or a lake nearby. I'll just fly down onto this tan stuff and look around.

**Carl**: Cathy, do you see something skinny and pink landing on the sand?

**Cathy**: Yes, but what could it be?

**Carl**: I've seen snakes and lizards, but I've never seen anything like this.

**Flossy**: Hi, green stubby plants. Could you please tell me where I am? Is there a good pond or lake around here? I am very thirsty and hungry.

**Carl & Cathy**: Ha, ha, ha!! A pond in a desert habitat! Pink lady, you must be kidding!

**Flossy**: I'm a flamingo, a beautiful wading bird from Florida. I need about three feet of water so I can scoop up my dinner.

**Cathy**: I am Cathy, a lovely cactus plant, and this is my brother, Carl. You don't look so beautiful to me. Those pink feathers are falling out.

**Carl**: You're SO skinny. Don't your parents feed you?

**Flossy**: Look, it's a long story. Just point me to where I can find a pond or a lake. I can't live without food and water.

**Carl & Cathy**: We don't need much water at all. We get food and water from our roots below ground. Then we store the food and water in our lovely stems. See?

**Flossy**: You're lucky. I've got to get out of this desert habitat right now. Which way is Florida?

**Carl & Cathy**: That-a-way.

**Narrator 4**: Carl and Cathy lean a little to the east.

**Carl & Cathy**: Goodbye and good luck. Don't forget to write.

**Flossy**: I'll need it! If only I can find my way back to my water habitat in Florida, I'll never fly away again.

### SCENE 5  *A shallow lake in Florida*

**Mother**: Look up, Dad. What is that flying zigzag in the sky?

**Freddy**: It's a flamingo, but it's very pale and skinny, not like anyone in our family.

**Father**: Hello. Come on down. We've got lots of plants and animals in our pond for you to eat.

**Flamingos**: Look, it can hardly walk. We'll splash some water on it.

**Flossy**: (whispering) Oh, thank you. Where am I?

**Mother**: You're in sunny Florida.

**Flossy**: Mom, Dad, don't you know me? It's Flossy. I'm so tired and hungry.

**Narrator 5**: Flossy zigzags into the lake and eats and drinks. She keeps scooping her bill into the water and mud.

**Flossy**: Oh, this tastes SO good! I'll never leave my habitat and my family again.

**Freddy**: I promise to share all of the food with you, Flossy. Don't worry. Your feathers will soon be bright pink again.

**Narrator 5**: So, the Flamingo family hugged at the edge of the lake and sang:

(To the tune of "Mary Had a Little Lamb")

Flossy had a habitat, habitat, habitat.
Flossy had a habitat,
Its lake was full of food.

She flew away from it one day,
It one day, it one day.
She flew away from it one day
'Cause Freddy was so rude.

She flew to other habitats, habitats, habitats.
She flew to other habitats,
She liked them NOT ONE BIT!

Then she found her habitat, habitat, habitat.
Then she found her habitat,
Where she can really FIT!

**Narrator 5**: So, Flossy Flamingo lived happily ever after and NEVER left her habitat again.

**Flossy**: Mom and Dad, what is writing?

**Mom and Dad**: Why would a flamingo want to write?

**Flossy**: It's a long story.

**THE END**

Primary Science Readers' Theatre

# ACTIVITIES FOR "HOW RACHEL RABBIT FIXED THE FOOD CHAIN"

## Prereading Activities

Students can enter their vocabulary into their Science Picture Dictionaries or the teacher can orally review the words with the class before reading the play.

camouflage   carnivores   consumers   herbivores

predator   prey   producers   nocturnal

Draw what you ate for dinner last night. Does this make you a consumer or a producer? Explain your answer.

## Purpose For Reading

Find out how the Rabbit family changed the food chain to solve a problem. As we read along, try to decide if you like their solution.

## Post-reading Activities

Write the problem and solution that the Rabbit family used in this play. Tell whether you agree or disagree with their solution. List at least two reasons for your answer.

Diagram an example of a real life food chain. Start with a producer (plant) and then draw the sequence of consumers until your food chain is completed. Label each step of the food chain.

Research an example of an herbivore and a carnivore. Find out what each one eats. Find out if the herbivore and carnivore are eaten by other animals. Use the Activity Sheet for *How Rachel Rabbit Fixed the Food Chain* to help you complete this activity.

Primary Science Readers' Theatre

# ACTIVITY SHEET
# "HOW RACHEL RABBIT FIXED THE FOOD CHAIN"

Name_____

Draw the herbivore and carnivore you researched. List what foods they eat and what animals eat them.

Herbivore

Carnivore

What do they eat?

_____

_____

_____

What do they eat?

_____

_____

_____

Who eats them?

_____

_____

_____

Who eats them?

_____

_____

_____

© 2002 Pieces of Learning

Primary Science Readers' Theatre

# "HOW RACHEL RABBIT FIXED THE FOOD CHAIN"
A play to teach about the food chains of animals

Characters
Mother Rabbit   Rapid Rabbit   Father Rabbit
Rachel Rabbit   Henrietta Hawk   Henry Hawk
Grass           Narrator        Samantha Snake
Sneaky Snake

Vocabulary  camouflage, carnivores, consumers, herbivores, predator, prey, producers, nocturnal

Props  rabbit ears, snakeskin shirts, hawk hats, green shirts (grass)

SCENE I        The prairie in the morning

**Narrator**: The Rabbit family is having a dinner meeting on a grassy hill. Father Rabbit is upset.

**Father**: Mother, Rapid, and Rachel, I have bad news. Your sister Rebecca Rabbit has been eaten by a hawk while she was eating her lunch on the prairie.

**Mother**: Oh, no! That's my fifth baby to be eaten this year! Between the hawks and the snakes, we'll become extinct! This will have to stop!

**Rachel**: How can it be stopped, Mother? We're herbivores; we feed on plants and grass.

**Rapid**: When we're eating, it's scary because a snake or hawk might sneak up and eat me.

**Mother**: I know because your five brothers and sisters have been eaten this year!

**Father**: I've thought of a plan. It means we'll have to change our habits, but if we stick to it, we'll all survive.

**Rachel and Rapid**: What's your plan, Father?

**Father**: Well, our fur is a good camouflage, but not good enough for hawks and snakes. I propose that we become nocturnal.

**Mother**: If we only come out at night, the hawks and snakes will be sleeping. Then we'll be safe.

**Rapid**: It's morning now, so let's go to sleep.

**Rachel**: I'm hungry, but I guess I can wait until dark.

**Narrator**: The Rabbits quickly hop into their hole.

**Rabbits**: Long live herbivores! Down with carnivores!

**Narrator**: The Rabbit family falls asleep.

### SCENE 2    *The prairie, a few days later*

**Narrator**: Henry and Henrietta Hawk are searching for food.

**Henry**: Henrietta, I'm hungry. I haven't seen any rabbits lately. Where could they be?

**Henrietta**: Beats me! Do rabbits take vacations?

**Henry**: If this keeps up, we'll have to eat those little rats and mice. Yuck!

**Henrietta**: We could eat the snakes, but they've been mighty skinny and tough lately.

**Narrator**: While Henry and Henrietta look for rabbits, the Rabbit family is sleeping snugly in their rabbit hole. Samantha Snake and Sneaky Snake slither in the grass, looking for something to eat.

**Samantha**: You're looking rather thin, Sneaky.

**Sneaky**: Well, I only have these little prairie mice to munch on. I haven't seen a rabbit in days!

**Samantha**: We carnivores need meat to eat. I wouldn't waste my time eating plants.

**Samantha and Sneaky**: We need a juicy, chubby bunny rabbit!

**Sneaky**: Look out for those hawks. Remember our dear sister Sally who was eaten by one just last week?

**Samantha**: I hope we find a rabbit soon. After all, we're the consumers.

**Sneaky**: Consumers?

**Samantha**: A consumer eats living things like plants and animals.

**Sneaky**: Goodie! I like being a consumer. Is a rabbit a consumer?

**Samantha**: Yes, because rabbits eat grass and other plants to survive.

**Sneaky**: Then is grass a consumer?

**Narrator**: Just then the Grass family speaks up.

**Grass**: We most certainly are NOT consumers! We are producers.

**Sneaky**: Producers?

**Grass**: We make our own food. We don't have to hunt for food. Our gorgeous green blades soak up the sunshine to make food.

**Narrator**: The hawks and snakes look puzzled. They wonder how the grass can make its own food.

**Grass**: We store our food in our lovely roots and leaves. It's too bad that all those herbivores eat us when we never bother anyone.

**Sneaky**: I see those hawks up there in the tree. Let's hide over there on the rocks. Our bodies will be camouflaged to protect us.

**Narrator**: As the rabbits sleep, the hawks and snakes look desperately for food.

SCENE 3    *The prairie, the next morning*

**Narrator**: Rachel Rabbit was so hungry, she snuck out of the rabbit hole while everyone was sleeping.

**Rachel**: (whispering) This nocturnal stuff is for owls, not rabbits. I need sunshine.

**Narrator**: Rachel is spotted by Sneaky Snake who quickly swallows her up.

**Samantha**: Sneaky, you didn't leave any for me!

**Sneaky**: It sure feels good to have a full stomach for a change.

**Narrator**: Just as quickly, Henrietta Hawk swoops down and gobbles up Sneaky Snake.

**Samantha**: Help! Help! It's a predator!

**Narrator**: All of the shouting woke up the Rabbit family. They hop outside to see what is happening.

**Rapid**: The sun is hurting my eyes!

**Mother**: We're not used to the sun anymore.

**Father**: Watch out for those hawks and snakes. We will NOT be their prey anymore!

**Mother**: Where is your sister, Rachel Rabbit?

**Rapid**: Come to think of it, I haven't seen her since last night.

**Narrator**: Samantha Snake sneaks up and tries to eat Rapid Rabbit, but he was too fast for her.

**Samantha**: You Rabbits make me sick! Your precious Rachel was eaten this morning by my brother, Sneaky Snake.

**Narrator**: Samantha begins to cry.

**Samantha**: Henrietta Hawk swooped down and ate HIM. Woe is me.

**Narrator**: Henry and Henrietta Hawk shout down from their tree.

**Hawks**: You Rabbits have upset the whole food chain on this prairie. We carnivores are starving!

**Grass**: We don't like being eaten at night by you Rabbits. We want to sleep and grow at night. Give us a break!

**Father**: Well, I guess I just can't change nature. We'll have to take our chances our here on the prairie. Poor Rachel.

**Mother**: Before we hide, we'd like to leave you with this song:

*(To the tune of "Pop Goes the Weasel")*

> All around the food chain,
> The snake ate the rabbit.
> Then the hawk ate the snake,
> Oh, my, what a habit!

> The consumers ate the producers,
> The predators ate their prey.
> That's the way the food chain goes,
> Day after day.

**Rabbits**: Let's hop on out of here before they get us!

### THE END

Primary Science Readers' Theatre

# ACTIVITIES FOR "THE PARTICLES VISIT ENERGYLAND"

## Prereading Activities

Students can enter their vocabulary into their Science Picture Dictionaries or the teacher can orally review the words with the class before reading the play.

    conductor    convection    energy    fluids    gases

    heat energy    liquids    particles    solids    temperature

Measure the temperature of your classroom. Put one thermometer on the floor and another as high up as possible. Wait a while and then record the two temperatures. Were they the same? Was one higher than the other? Try to explain why they are not the same.

## Purpose For Reading

Find out why the temperature is higher near the ceiling of our classroom.

## Post-reading Activities

Look back at the reasons you wrote for the temperature differences in our classroom. Were you correct? Think about what happened to the Particles when Heat Energy turned on the electric heater. If you need to change your answer, please do.

Use the Activity Sheet for *The Particles Visit Energyland* to show which form of matter you would prefer if YOU were a Particle.

Draw an example of convection in a liquid and convection in a gas. Put dots in for the particles and use arrows to show their movement. Explain what is happening to the particles.

Primary Science Readers' Theatre

# ACTIVITY SHEET
# "THE PARTICLES VISIT ENERGYLAND"

Name _____

Decide whether you would rather be a Particle in a solid, a liquid or a gas that is being heated. Think about which solid, liquid or gas you would like to be part of. Draw a picture of yourself in this form of matter. Then compose a story telling about your adventure. Remember to give reasons why you like being there.

I would like to be a Particle in _____ that is being

heated because _____

_____

_____

It's fun because _____

_____

_____

_____

_____

_____

_____

_____

© 2002 Pieces of Learning

Primary Science Readers' Theatre

# "THE PARTICLES VISIT ENERGYLAND"
A play to teach how energy moves through matter

Characters

| | | | | |
|---|---|---|---|---|
| Narrator | Mr. Cook | Polly Particle | Peter Particle | Mom |
| Particles | Dad Particle | Heat Energy | | |

Vocabulary  conductor, convection, energy, fluids, gases, heat energy, liquids, particles, solids, temperature

Props  Cut out capital "H" and "P" letters for students to tape to their shirts, an energy wand, a paper flame

*Scene 1          The kitchen of the Cook's house*

**Narrator:** Mr. Cook is frying eggs for breakfast.

**Mr. Cook:** I'll just turn on the gas burner and fry these eggs. I'm in a hurry, so I'll use a high flame.

**Heat Energy:** That Particle family has been sleeping. Wait until my energy wakes them up!

**Particles:** Whoa! Who's bumping me? I was taking a nap and minding my own business! It's getting awfully hot!

**Polly:** Last thing I remember, we were inside the eggs in the refrigerator.

**Mom:** We've been living inside the eggs. We had plenty of room inside their cool liquid.

**Dad:** Now I feel more solid, and I'm getting jumpy.

© 2002 Pieces of Learning

**Heat Energy**: Yes, Mr. and Mrs. Particle, Polly and Peter. Heat energy can change you Particles from a liquid to a solid.

**Peter**: Stop bumping into me, Polly!

**Polly**: I can't help it. I keep dancing in this frying pan. I am jumping and hopping. I can't stop!

**Particles**: Wow! We never moved so fast in our lives!

**Heat Energy**: As you Particles get hotter, you will move much faster. Calm down. You Particles are tiny parts of matter. I have the power to move you around.

**Mr. Cook**: It looks like the eggs are ready. I'm hungry.

**Particles**: Mr. Cook is going to eat us! We don't want to be eaten!

**Heat Energy**: Yes, it's true, but with my magic energy wand, I can stop time and show you around Energyland.

**Particles**: Oh, please take us to Energyland.

### SCENE 2  Energyland, U.S.A.

**Narrator**: The Particles and Heat Energy ride the Energy tram.

**Heat Energy**: The first stop is Solid Matter. This is where you Particles hop inside that spoon.

**Particles**: Why should we do that?

**Heat Energy**: A spoon is a solid. You're going to go for a ride. Hang on! It will be fun.

**Particles**: I hope it's better than being in the frying pan.

**Narrator**: The Particles get off the tram and jump into a spoon. The spoon is placed in a cup of very hot water.

**Particles**: I'm spinning!

**Polly**: Peter, stop bumping me.

**Peter**: I'm not doing it on purpose.

**Heat Energy**: Polly, he's right. Now that you are part of the spoon, a solid matter in a hot liquid, you will move faster and bump into each other.

**Particles**: We're SO close together.

**Heat Energy**: That's the way heat energy moves through solid matter like a spoon. The spoon is metal. Metal is a good conductor of heat energy. I can travel through a spoon before you can say "conductor."

**Mom and Dad**: We're tired of moving so fast. Can we visit another kind of matter? We need a rest.

**Heat Energy**: Had enough so soon? This ride is like bumper cars. O.K. Hop off, and we'll go to Liquid Matter.

**Narrator**: Heat Energy and the Particles ride the Energy tram to Liquid Matter.

**Heat Energy**: Jump into this pot of hot chocolate. I'm just about to start heating it up.

**Narrator**: The Particles dive into the pot as Heat Energy turns on his flame under the pot.

**Particles**: Ah, this feels much better. It's nice and cool. We're moving around slowly in the liquid.

**Polly & Peter**: It's nice here at the bottom of the pot.

**Mom & Dad**: It's very relaxing. Not like that Solid Matter, thank goodness.

**Heat Energy**: I'm turning up my energy and increasing the temperature. Here we go!

**Polly**: Oh, my! What's happening? I'm spinning and getting farther away from the rest of you. It's getting hot, and I'm feeling lighter now.

**Peter**: Me, too, Polly. I'm moving up now. Follow me!

**Narrator**: As the fluid becomes hotter, Peter and Polly Particle get warmer and lighter. They move upward to the top of the pot of hot chocolate.

**Mom & Dad**: Look! Peter and Polly are moving up. Why can't we?

**Heat Energy**: Don't worry. You will move up as soon as you get hot enough.

**Mom & Dad**: Great. Here we go!

**Heat Energy**: This circular movement is called convection.

**Particles**: We like this much better than Solid Matter. In Liquid Matter, we are farther apart, and we rise to the top when we get warm. It's better than all that bumping and crashing into each other.

**Heat Energy**: Ready for another ride?

**Particles**: Sure! Where to next?

**Heat Energy**: Hop onto the Energy tram. Next stop is Gas Matter.

**Narrator**: The Particles and Heat Energy ride the tram to Gas Matter. They enter a large room.

**Heat Energy**: Hop off the Energy Tram and rest awhile on the floor.

**Peter**: Cool! Spread out!

**Polly**: I'm trying, Peter, but we are much closer together than we were in the hot chocolate.

**Heat Energy**: That's because it's cool on the floor. Wait until I turn up the temperature. Here we go!

**Narrator**: Heat Energy turns on a large electric heater. As the air gets warmer, the Particles move faster and move apart from each other.

**Mom and Dad**: Good! We're less crowded, and we're rising up into the air!

**Polly**: I feel light as a feather!

**Peter**: Me, too! It's like going up in a hot air balloon.

**Mom and Dad**: Now that we're near the ceiling, we aren't getting as much heat energy. We're feeling drafty.

**Heat Energy**: That's right! As you Particles lose heat, you will move closer together and slow down. You'll start to sink.

**Particles**: Down we go! This is like a roller coaster. Up and down and up and down again!

**Heat Energy**: As you come near me again, you'll heat up and rise again.

**Particles**: Weeee!!! We like this ride!

**Heat Energy**: I see you like convection. That's what it's called when particles are heated in gas or liquid. They rise to the top and then cool off and come back down and do it all over again.

**Particles**: We'd like to stay here forever!

**Heat Energy**: Sorry, but our time is up. We have to go back to Mr. Cook's kitchen.

**Particles**: Oh, no! We'll be eaten.

**Narrator**: The Particles and Heat Energy get in the Energy Tram and ride back to Mr. Cook's house.

SCENE 3     Mr. Cook's kitchen

**Narrator**: The Particles sadly climb back into the egg in the frying pan.

**Heat Energy**: Well, I hope you enjoyed your trip to Energyland.

**Polly**: We did, Heat Energy, but what will become of us now?

**Heat Energy**: Particles, it's time for Mr. Cook to eat his eggs for breakfast. Sorry it has to end this way, but I did find you living inside the eggs.

**Peter**: All right, Heat Energy, but please stop time for just one more minute so we can sing you a song.

*(To the tune of "London Bridge Is Falling Down")*

Solid Matter, heating up,
Heating up, heating up.
Solid Matter, heating up,
Bump go the Particles.

Liquid Matter, heating up,
Heating up, heating up.
Liquid Matter, heating up,
Up go the Particles.

Gas Matter, heating up,
Heating up, heating up.
Gas Matter heating up,
Up go the Particles.

Now you see what heat can do,
Heat can do, heat can do.
Now you see what heat can do,
It's the power of energy!

**Particles**: So long, Heat Energy. It was great learning about conductors and convection.

**Heat Energy**: Glad you enjoyed the ride, Particles.

**Narrator**: Heat Energy leaves, and Mr. Cook takes the eggs out of the frying pan and puts them onto a plate.

**Mr. Cook**: I think I'll use a little salt and pepper. Yum, yum. Delicious!

## THE END

# ACTIVITIES FOR "STOP! YOU'RE INTERRUPTING MY LIFE CYCLE!"

## Prereading Activities

Students can enter their vocabulary into their Science Picture Dictionaries or the teacher can orally review the words with the class before reading the play.

adult   characteristics   embryos   germinate   incubating

life cycle   migrated   seed coat   seedling   traits

Ask the class if they are adults. Discuss and list what stages people go through to become adults. Explain that plants go through stages before they are fully grown, and that keeping this in mind should help them to better understand the play.

## Purpose For Reading

Robbie and Roberta Robin are hungry. They find some newly planted tomato seeds. Read to find out why they decide NOT to eat the tomato seeds.

## Post-reading Activities

Do you agree or disagree with Robbie and Roberta Robin's decision NOT to eat the tomato seeds? List two reasons for your answer.

Draw and label the life cycle of a robin and the life cycle a tomato. Write about how long each life cycle lasts.

Research the life cycle of another plant or animal. Draw a diagram of the life cycle and label the different stages. Compare its life cycle to the robin's or the tomato's. Use the Activity Sheet for *Stop! You're Interrupting My Life Cycle!* to help you.

Primary Science Readers' Theatre

# ACTIVITY SHEET
# "STOP! YOU'RE INTERRUPTING MY LIFE CYCLE!"

Name _____

Draw the stages of the life cycle of the plant or animal you have researched. Label each stage of its cycle.

**THE LIFE CYCLE OF** _____

© 2002 Pieces of Learning

Primary Science Readers' Theatre

> Now, let's compare the life cycle of that plant or animal to a robin's or a tomato's life cycle. Complete the sentence below. Then tell how their life cycles are similar and different.

I am comparing the life cycle of a _____ to the life cycle of a _____.

| How they are similar | How they are different |
|---|---|
| 1. _____ | 1. _____ |
| 2. _____ | 2. _____ |
| 3. _____ | 3. _____ |
| 4. _____ | 4. _____ |

© 2002 Pieces of Learning

Primary Science Readers' Theatre

# "STOP! YOU'RE INTERRUPTING MY LIFE CYCLE"
### A play to teach the life cycles of plants and animals

Characters
Narrator          Mother Robin      Father Robin      Robbie Robin
Roberta Robin     tomato seeds

Vocabulary  adult, characteristics, embryos, germinate, life cycle, incubating, migrated, seed coat, seedling, traits

SCENE 1          Early spring in Birdland, New Jersey

**Narrator**: Mother and Father Robin are having a chat with one-year-old Robbie and Roberta Robin.

**Father**: Robbie and Roberta, you probably don't remember that you were hatched in a nest in this very tree just one year ago.

**Robbie & Roberta**: Is it our birthday?

**Mother**: Yes, happy birthday. Here's a yummy worm for each of you.

**Father**: Your mother worked very hard incubating you when you were just little blue eggs. It took twelve days before you both hatched.

**Robbie**: I remember how warm and cozy it felt inside the egg.

**Roberta**: I remember my egg being turned over a lot. I was getting dizzy!

**Mother**: Yes, I turned you over to warm you evenly. I wanted you to hatch!

**Father**: You were just little white spots inside the yolk until you began to grow inside your eggs.

**Mother**: You were called embryos until you hatched.

**Narrator**: Now, a year later, Roberta and Robbie look very much like their parents.

**Mother**: Roberta and Robbie, you sure look like your Dad!

**Father**: You both sound like your mother.

**Robbie & Roberta**: We know. Those are traits or characteristics that we got from you. Most of our brothers and sisters look like us and act like us, too.

**Mother**: So, it's been a year, and you both are in the adult stage of your life cycle.

**Robbie**: Life cycle? Is that like a bicycle?

**Roberta**: Very silly, Robbie. First, we were embryos inside our warm eggs. Then Mom sat on us and incubated us until we hatched.

**Father**: Now you are fully grown adults.

**Mother**: I remember that you two were SO hungry. We fed you 80 or 90 times a day!

**Father**: I was very busy finding juicy worms and insects for you both. And you sure loved eating cherries, grapes and tomatoes.

**Mother**: Last fall we migrated South for the winter.

**Father**: I'm glad to be back in Birdland, New Jersey. Now Mother can lay more eggs and hatch more baby robins.

**Robbie & Roberta**: More robins! We already have 30 brothers and sisters!

**Mother**: Yes, and now that you are grown, you can hunt your own food while I sit on the new eggs in my nest.

**Robbie & Roberta**: Bummer! Now we have to find our own food.

**Roberta**: Well, let's not waste any time, Robbie. I'm hungry.

**Robbie**: Me, too. We better get going. See you later, Mom and Dad.

**Mother**: Good luck! Dad, let's get busy making the nest.

**Dad**: We'll need some grass, twigs, and mud.

**Narrator**: Robbie and Roberta Robin fly off together while Mother and Father Robin build a new nest.

SCENE 2      *Miss Tulip's garden, Birdland, New Jersey*

**Narrator**: Robbie and Roberta Robin see Miss Tulip hoeing and planting seeds in her garden. They sit in a tree and wait until she is finished planting.

**Roberta**: Robbie, that looks like a good spot to start searching for worms.

**Robbie**: The ground is soft, so we can peck around for our dinner.

**Roberta**: If you see any insects, open your mouth wide, and grab them before they know what hit them!

**Narrator**: Robbie and Roberta dig under the earth with their beaks. They are pecking where Miss Tulip just planted tomato seeds. Just as Robbie pecked at a tomato seed, he heard a lot of noise and chatter.

**Seeds**: Hey, just what do you think you're doing interrupting our life cycle?

**Robbie & Roberta**: Life cycle? You tomato seeds have a life cycle?

**Seeds**: Yes, we may be tiny, but if you'll stop chomping on our seed coats, we will start to grow.

**Robbie**: What will you become?

**Seeds**: Well, if we have a complete life cycle, like you had, we will sprout or germinate into embryos.

**Roberta**: We were embryos last year. We lived inside little blue eggs, and our Mother sat on us to keep us warm until we hatched.

**Seeds**: Well, Miss Tulip planted us underground and watered us. We need to be warm and moist to germinate.

**Robbie & Roberta**: How long will that take? Can we eat you then?

**Seeds**: It will take many weeks for us to sprout and push up out of the soil to become seedlings.

**Robbie & Roberta**: Seedlings?

**Seeds**: Yes, little plants. Then, if we get lots of air, water and sunlight, we will grow very tall. Then we grow little yellow flowers on our stems.

**Robbie**: Can we eat you then?

**Seeds**: Not if you want us to have a complete life cycle. You'll have to wait until July. Then it will be hot, and the flowers will turn into green baby tomatoes. We grow and grow until our tomatoes ripen. We'll be big, red and juicy by summer.

**Robbie & Roberta**: Then can we eat you?

**Seeds**: Yes. When we are fully grown, we will be delicious and ready to eat.

**Roberta**: OK. We'll cover you up with this soil and come back in July.

**Seeds**: Goodbye, robins!

**Robbie & Roberta**: Goodbye, tomato seeds.

**Robbie**: It looks like we can't get any fruit or tomatoes for a few months. It's too early, so we'll have to hunt for worms and insects.

**Narrator**: Roberta and Robbie Robin fly to the lawn of Birdland Elementary School. They peck and peck until they find many juicy worms.

**Roberta**: Yum, yum! I sure was hungry!

**Robbie**: Me, too. Let's fly back home and tell Mom and Dad.

**Narrator**: The Robins fly back to the maple tree.

## SCENE 3    *The maple tree*

**Narrator**: Roberta and Robbie find Mother and Father Robin finishing a beautiful nest.

**Robbie**: Looks like you've been busy.

**Mother**: How did you make out? Did you find any food?

**Roberta**: Yes. First we found tomato seeds in Miss Tulip's garden. They insisted that we leave them alone so they could have a chance to germinate.

**Father**: Yes, plants have life cycles, too, Roberta and Robbie.

**Robbie**: They won't be ready for us to eat until July. So, we dug up some worms on the school lawn.

**Mother**: Don't forget to look for insects, too. They are delicious.

**Father**: It looks like you two did well for yourselves today.

**Robbie & Roberta**: Our bellies are full, and we're glad to be back home.

**Mother**: Soon you'll each start a family of your own.

**Robbie**: Is that part of the life cycle?

**Father**: It sure is! Then Mother and I will be grandparents.

**Roberta**: This life cycle stuff is getting me tired. I need a rest. But before I birdnap, I have a song for you.

*(To the tune of "Did You Ever See A Lassie?")*

Did you ever see an embryo, an embryo, an embryo?
    Did you ever see an embryo hatch from an egg?
Did you ever see an embryo, an embryo, an embryo?
    Did you ever see an embryo hatch into a bird?
Grow this way and that way and this way and that way.
    Did you ever see an embryo hatch into a bird?

Did you ever see an embryo, an embryo, an embryo?
    Did you ever see an embryo sprout from a seed?
Did you ever see an embryo, an embryo, an embryo?
    Did you ever see an embryo grow into a plant?
Grow this way and that way and this way and that way.
    Did you ever see an embryo become an adult?

**Robbie**: This hunting for my own food is exhausting! I'm going to birdnap.

**Mother and Father**: Sweet dreams!

**Narrator**: Robbie and Roberta Robin fall asleep and dream about juicy tomatoes.

**THE END**

# ACTIVITIES FOR "THE SUNBEAMS TAKE A TRIP"

## Prereading Activities

Students can enter their vocabulary into their Science Picture Dictionaries or the teacher can orally review the words with the class before reading the play.

    indigo    lightning    natural light    prism    reflected light

With a partner, list all of the things that give us light on Earth. Make a class list and discuss the answers.

## Purpose For Reading

Sunny and Sammy Sunbeam are bored with their daily job of beaming. Their parents send them on a trip to Earth. Predict what will happen to them on their trip.

## Post-reading Activities

Go back to the list of things that give off light. Ask students to add to the list. Then have the students categorize the sources of light into manmade or natural light.

Ask students to write a story telling what would happen if the sunbeams took a day off. Students can illustrate their stories.

Use a prism to bend white light. Draw a rainbow and label the colors you saw. Use the Activity Sheet for *The Sunbeams Take a Trip* to help you.

Primary Science Readers' Theatre

# ACTIVITY SHEET
# "THE SUNBEAMS TAKE A TRIP"

Name _____

Draw a picture of a rainbow. Show all of the colors in order and label each color.



To see a rainbow, there needs to be _____

_____

_____

© 2002 Pieces of Learning

Primary Science Readers' Theatre

# "THE SUNBEAMS TAKE A TRIP"
A play to teach about light and color

Characters
Sunny Sunbeam      Sammy Sunbeam      Mom Sunbeam
Dad Sunbeam        Fireflies          Fanny Firefly
Lightning          Moonbeams          Narrator
Luke Lightning

Background  outer space with the sun, earth and moon, a rainbow

Costumes  yellow shirts or sun, moon, firefly and lightning symbols

Props  flashlight, light bulb, prism

Vocabulary  indigo, lightning, natural light, prism, reflected light

SCENE 1     On the sun

**Narrator**: The Sunbeam family is busy working. They are spreading their sunshine to all of the universe.

**Dad**: Looks like another sunny day today.

**Mom**: It's SO hot! I wish just once that we could have a cool breeze.

**Sunny**: It's the same boring thing every day. Get up, shine my beam, sleep, get up, shine my beam.

**Sammy**: Me, too. I'm tired of just being a sunbeam, day in and day out.

**Sammy & Sunny**: We need a break!

© 2002 Pieces of Learning

**Mom**: Well, I do have some relatives you could visit, but you have to be on your best behavior.

**Sunny & Sammy**: We will! We will!

**Dad**: I'll pack you a snack to take along on your trip.

**Mom and Dad**: Be careful! Mind your manners!

**Sunny & Sammy**: Bye! See you in a few days.

**Narrator**: Mom and Dad wave as Sunny and Sammy leave the sun.

*SCENE 2     On the earth*

**Narrator**: Sunny and Sammy travel along their sunbeams to earth.

**Sammy**: Sunny, it is taking much longer than I thought to get to earth.

**Sunny**: Well, Sammy, the sun is 93 million miles away from the earth, so it is going to be a long trip.

**Narrator**: Sunny and Sammy finally arrive at the home of their cousins, the Fireflies.

**Fireflies**: We have been expecting you. Your mom and dad beamed us that you were coming.

**Fanny F**: You're much bigger than I thought you would be.

**Sunny**: We look smaller when we are far away, Fanny.

**Fireflies**: It's getting so hot in here with you two sunbeams.

**Sammy**: We are just doing our job, sending out light and heat.

**Sunny and Sammy**: Mom and Dad told us to be good little sunbeams.

**Fanny**: We're only little fireflies, so we are not as bright as you.

**Narrator**: Suddenly the Firefly house begins to shake. There is a loud bang.

**Sunny**: Who's that?

**Sammy**: This earth place is scary!

**Fireflies**: Oh, don't worry. It's just thunder and lightning.

**Fanny**: We told Luke Lightning and his family to come and meet their cousins.

**Narrator**: The Fireflies open the door as bright flashes of lightning come inside.

**Luke**: Hello, cousin Fanny and all of you Fireflies.

**Lightning**: You are looking nice and bright today.

**Fanny**: My Lightning cousins are looking very flashy today.

**3 Fireflies**: We'd like you to meet Sunny and Sammy Sunbeam. They have come all the way from the sun for a visit.

**Sunny & Sammy**: We would shake hands, but we don't have any.

**Sunny**: Cousin Luke, what is that white thing up there on the ceiling?

**Luke**: I don't know. Cousin Fireflies, what is that white thing up there?

**Fireflies**: It's an electric light bulb. We use it for extra light.

**Lightning**: We don't need light bulbs. We are so large and bright.

**Sunny & Sammy**: Our glow is so bright. We don't need light bulbs on the sun, either.

**Fireflies**: On earth we have natural light and light made by people.

**Sunny & Sammy**: People? What are people?

**Fanny**: See those weird creatures walking outside?

**Lightning**: Those things are called people.

**Fireflies**: On earth, we have lots of light in the daytime from sunbeams.

**Lightning**: Can we give you some of our light?

**Fireflies**: No thanks. That's too powerful for us!

**Fanny**: Anyway, these earth people have figured out how to make their own light.

**Sunny & Sammy**: We thought we were doing our jobs right.

**Lightning**: You are, but at night we still need light.

**Luke**: So earth people invented lots of ways to make light and heat.

**Fanny**: We have fire and matches to light candles.

**Fireflies**: We have lights and lamps that use oil, gas or electricity.

**Lightning**: Those earth people even made flashlights that run on batteries.

**Sammy**: Light bulbs? Flashlights? Fire? Matches?

**Sunny**: Earth people must be very smart!

**Sunny & Sammy**: Cousins, what is that wet stuff falling from the sky?

**Fireflies & Lightning**: It's called rain.

**Narrator**: Just then, the rain stops. The sun comes out, and a huge rainbow streaks across the sky.

**Sunny**: What is that bright thing in the sky?

**Lightning & Fireflies**: It's a rainbow!

**Sammy**: We don't have rain or rainbows on the sun.

**Luke**: Didn't you know that your light only looks white?

**Fanny**: Inside your beams there are beautiful colors!

**Lightning & Fireflies**: Red, orange, yellow, green, blue, indigo and violet.

**Sunny & Sammy**: Mom and Dad always told us to send our beams in a straight line.

**All cousins**: If your beams of light are bent by passing through raindrops or a prism, we can see all of your beautiful colors.

**Sunny**: I never knew we were so beautiful!

**Sammy**: Thanks for showing us around the earth. It's time for us to finish our trip.

**Sunny & Sammy**: It was super meeting our earth cousins.

**Narrator**: The sunbeams wave as they leave earth.

## SCENE 3    On the moon

**Narrator**: Sunny and Sammy beam off to the moon to meet their cousins, the Moonbeams.

**Sunny**: Sammy, so far we have used our best manners.

**Sammy**: Let's keep it up. It's not too far to the moon. I see them waving to us.

**Moonbeams**: Welcome, Sunny and Sammy Sunbeam. Did you enjoy your visit to the earth?

**Sunny**: Yes. Did you know the earth people can make their own light?

**Moonbeams**: We don't need to make extra light. We have such strong moonbeams.

**Sunny and Sammy**: Ha, ha, ha!

**Sammy**: Did you Moonbeams know that your light comes from the sun?

**Moonbeams**: That's a mean thing to say!

**Sunny**: Oh, dear. We told Mom and Dad we would use our best manners and now we've upset the Moonbeams!

**Sammy**: Cousin Moonbeams, your light is reflected from sunlight.

**Moonbeams**: How dare you say that! Moonbeams are very bright and strong.

**Sunny**: You see, the sun is a huge star. We send you our sunbeams. Then they are reflected back.

**Moonbeams**: Reflected? What does that mean?

**Sunny & Sammy**: Well, we do our jobs all day. We beam straight down on the earth and the moon.

**Moonbeams**: So what?

**Sunny**: Then our light bounces off the moon. It looks like it's moonlight, but it really is sunlight. It hits the moon and bounces off of it. That is reflected light.

**Moonbeams**: All this time we thought we were being good moonbeams.

**Sammy**: You were. You just didn't know that we were doing all the work.

**Moonbeams**: That's true. We're never tired. We always have time to play moonball.

**Sunny**: That's why we took this trip. Sammy and I were tired of working all day. It's hard to send your best beams out every day.

**Moonbeams**: Oh, thank you, Sunny and Sammy Sunbeam.

**Sammy**: We are so happy that you aren't mad at us anymore.

**Moonbeams**: How could we be angry when you are the reason we have light on the moon? We are so thankful.

**Narrator**: Sunny and Sammy are feeling homesick. They want to get back to Mom and Dad Sunbeam soon.

**Sunny & Sammy**: We had fun visiting the moon and meeting our Moonbeam cousins, but it is time for us to go back home now.

**Moonbeams**: So long, cousins. Keep sharing your light with us!

**Narrator**: All wave goodbye, and Sammy and Sunny zoom off.

## SCENE 4    Back on the sun

**Narrator**: Mom and Dad wave as Sunny and Sammy return.

**Mom**: How was your trip?

**Sunny**: We like our earth and moon cousins.

**Sammy**: We used our best manners.

**Dad**: That's good! Did you learn anything new?

**Sunny & Sammy**: Boy, did we!

**Sammy**: We learned that earth creatures can make their own light when sunlight isn't enough for them.

**Mom**: How are the Moonbeams?

**Sunny**: They were surprised to find out that the moon reflects sunlight.

**Dad**: You must be tired after that long trip.

**Sunny & Sammy**: Yes, but first we have a song for you.
*(To the tune of "Yankee Doodle")*

Sunny and Sammy left the sun,
Riding on their sunbeams.
Met their cousins on the earth
And on the moon met moonbeams.

Sunny and Sammy, shine your light,
So we can have good weather.
Keep us warm and keep us bright,
Please shine your beams forever.

**Narrator**: The sunbeams all hug each other.

**Sammy**: Let's rest up for tomorrow, Sunny. We have a big job to do!

**Sunny**: You bet!    **THE END**

# ACTIVITIES FOR "WALLY THE WATER WASTER"

## Prereading Activities

Students can enter their vocabulary into their Science Picture Dictionaries or the teacher can orally review the words with the class before reading the play.

    aquifer   desalination   reservoir   surface water   vapor

Ask the class what would happen if people wasted so much water that none was left on earth? (The teacher can list the possibilities on chart paper.)

## Purpose For Reading

Find out how Wally the Water Waster feels about wasting water after his visit from the Water Savers.

## Post-reading Activities

Work with a partner. Write two questions you would ask Wally about water. Write answers to your questions as if you were really Wally. Decide which one of you will be Wally and which one of you will be the interviewer. Practice the questions and answers orally. When everyone is ready, the class can listen to the interviews.

Research the source of your town's water. Compare your city's water source to the one in Smalltown, U.S.A. Tell how they are alike and how they are different.

Did you know that each person uses between 80-100 gallons of water a day? Using the Activity Sheets for *Wally the Water Waster,* total how much water you used each day for one week. Make a class line graph or bar graph to compare the total water used for each water use (toilet, hand washing, bath, shower, face washing, tooth brushing). Another graph can compare each student's total water use for the week.

# ACTIVITY SHEET
# "WALLY THE WATER WASTER"

Name_____ Week of_____

Tally your use of water at home and then use the Water Usage Chart to total how much water you use in one week. Hang this paper near the bathroom.

TOILET - Time used _____

Total water used from flushing_____

HAND WASHING - Time used _____

Total water used from hand washing_____

FACE WASHING - Time used _____

Total water used from face washing_____

SHOWER - Time used _____

Total water used from showers_____

BATH - Time used _____

Total water used from baths_____

TOOTH BRUSHING - Time used _____

Total water used from brushing teeth_____

The total water I used this week was _____ gallons.

I could use less water if I _____

Primary Science Readers' Theatre

## WATER USAGE LIST

**Shower**      20 gallons

**Bath**        30 gallons

**Toilet flush**   1 and ½ gallons   (Toilets more than 10 years old use 3 gallons.)

**Face washing**   2 gallons

**Hand washing**   2 gallons

**Tooth brushing**  1 gallon

<u>How much water would you use in a month?</u>

**Shower** _____ gallons        **Toilet flush** _____ gallons

**Bath** _____ gallons          **Face washing** _____ gallons

**Hand washing** _____ gallons  **Tooth brushing** _____ gallons

<u>How much water would you use in a year?</u>

**Shower** _____ gallons        **Toilet flush** _____ gallons

**Bath** _____ gallons          **Face washing** _____ gallons

**Hand washing** _____ gallons  **Tooth brushing** _____ gallons

© 2002 Pieces of Learning

Primary Science Readers' Theatre

# "WALLY THE WATER WASTER"
A play to teach the sources of water and the importance of water conservation

Characters
Wally    O    H-1    Narrator    H-2

Vocabulary  aquifer, desalination, reservoir, surface water, vapor

Scenery  can show a town by a river, a river and dam with a reservoir, a town by the ocean

Costume  Wally wears a t-shirt and shorts
H-1, H-2, and O wear blue shirts with letters H or O
The class will be H-1, H-2, and O

SCENE 1    Smalltown, U.S.A.

**Narrator**: One lazy, hazy summer afternoon, Wally the Water Waster was washing the family car. He got tired and decided to take a nap under the maple tree. Just as he was falling asleep, he heard strange voices.

**H-1**: Hey, buddy, you forgot to turn off the hose. That's wasting water!

**Narrator**: Wally the Water Waster woke up and opened his eyes. He saw three little blue creatures standing over him. Two had the letter H on their shirts and the other had a big O.

**Wally**: Wh-wh-who are you? Am I dreaming or is this a nightmare?

**H-2**: Wally, we are the Water Savers. Water is made from two parts hydrogen to one part oxygen. So, we make $H_2O$ or water.

© 2002 Pieces of Learning

**O**: We see that you have been wasting earth's valuable water today. You forgot to turn off the hose when you took a nap.

**Wally**: I didn't forget. I was planning to finish washing the car in a little while. I just left the water on because I wasn't finished.

**H-1**: But Wally, if everyone did that, we might run out of water. I'll just turn off this hose.

**Wally**: Oh, come on. Every time I turn on the faucet, I always get water.

**H-2**: Yes, Wally, that's because Smalltown is lucky to be next to a river. Water from the river is pumped out, sent to a treatment plant where it is cleaned, and then sent to a pumping station.

**Wally**: I never saw a pumping station in Smalltown.

**H-1**: It's near the river on Elm Street.

**O**: Then the clean water is pumped through pipes to fire hydrants and to all the houses here in Smalltown.

**Wally**: So, why did you turn off the hose?

**H-1, H-2, O**: Why don't you come with us in our magic blue watercopter and we'll show you?

**Narrator**: Wally climbs into the watercopter along with the Water Savers. O is the pilot. They fly above Smalltown's river and head west.

## SCENE 2    *The Hoover Dam, Arizona*

**Narrator**: As the watercopter flies over the Rocky Mountains, Wally sees a huge lake and a cement wall.

**Wally**: H-1, what is that over there? I never saw anything like that in Smalltown!

**H-1**: That's a manmade lake or reservoir in the Colorado River.

**Wally**: How did it get so big? Why is that giant wall there?

**H-2**: Lots of big cities like Los Angeles, California, need fresh water for their houses and businesses.

**O**: So, they built the Hoover Dam to hold back the river water. Then it is sent through pipes all the way to California.

**Wally**: I guess Los Angeles doesn't have a river nearby like we do in Smalltown.

**H-2**: You're right. The people in Smalltown are lucky to have surface water nearby. Surface water is the easiest and cheapest source of water.

**Wally**: But I still don't see why running the hose a lot is a big deal.

**O**: Well, Wally, the earth's surface is 3/4 water, but it's mostly salt water. We can't drink it or use it for cooking or farming.

**H-2**: Fresh water is only 3% of the earth's water and 2/3 of that is frozen.

**Wally**: That leaves only 1% for us to use!

**O**: Wally, did you know that the average person in the United States uses 75 gallons of water every day?

**Wally**: Seventy-five gallons a day?

**H-1**: Let's fly in the watercopter, and we'll show you another way that people find fresh water.

**Narrator**: So, the Water Savers and Wally fly southeast to Miami, Florida.

### SCENE 3    *Miami, Florida*

**Narrator:** The watercopter lands on the beach.

**Wally:** It's too bad I don't see a lake or river nearby. I see the Atlantic Ocean, but now I know that all that salt water cannot be used.

**O:** Miami uses underground water for its supply of fresh water.

**Wally:** Underground water? How do they do that?

**H-2:** Wells are drilled to find a layer of rock where fresh water collects.

**H-1:** They are called aquifers.

**Wally:** Aquifers? How do they get the water out of the aquifers?

**Narrator:** $H_2O$ explains that windmills or electric pumps are used to pump fresh water from the underground aquifers to the treatment plant.

**Wally:** That must cost a lot of money to do all of that digging and pumping!

**H-1:** That's another reason why we have to conserve our fresh water. When you kept the hose running, you were wasting all of that fresh water!

**O:** Now, Wally, let's show you a place where there is no surface fresh water OR underground fresh water.

**Narrator:** So, Wally and the Water Savers fly west in the watercopter.

### SCENE 4    *Santa Barbara, California*

**Narrator:** The watercopter lands on the beach in California.

**Wally:** Where are we? Is this the Atlantic Ocean?

**O:** No, Wally, this is the Pacific Ocean. We are now in Santa Barbara, California.

**H-2**: They don't have a lake or river or an aquifer.

**Wally**: Do they have a dam or a reservoir? How do they get their fresh water?

**H-1**: They have to use the salt water from the ocean. A special place called a desalination plant removes the salt from the ocean water.

**O**: The ocean water is heated and evaporated. Then the water changes to water vapor.

**H-2**: The salt is removed, and then the water vapor is cooled and turned back into water.

**Wally**: So now it's fresh water. Does it cost a lot to desalinate ocean water?

**H-1, H-2, O**: You bet it does!

**Narrator**: Wally and the Water Savers leave Santa Barbara, California, and fly back to Smalltown, U.S.A.

### SCENE 5    *Smalltown, U.S.A.*

**Narrator**: On the way back home, Wally notices a river.

**Wally**: Look down below. The river in Smalltown isn't dirty and dark like that river.

**O**: Do you see all of the factories nearby?

**H-2**: Many of them have been dumping waste and chemicals into the river.

**Wally**: Yuck! And people drink that water?

**H-1**: Yes, but first it goes to the water treatment plant. The water is cleaned there and checked for bacteria.

**H-2**: Sometimes chemicals ruin fresh water, and even the water treatment plants can't clean the water.

**Wally**: I never realized there were so many problems with fresh water in the United States. Is it like this all over the world?

**O**: Yes. It's even worse in countries where there is very little rain.

**H-2**: In some places, people have to travel far to find fresh water.

**H-1**: Some people even have to carry big jugs or pitchers of water from wells back to their homes.

**Wally**: Now I see why you told me to turn off the hose!

**Narrator**: The watercopter lands at the Smalltown Airport.

**Narrator**: Wally and the Water Savers sit down under the maple tree in his backyard.

**H-1, H-2, O**: It's almost time for us to leave, Wally, but we have a little song to help you remember what you learned today.

*(To the tune of "Row, Row, Row Your Boat")* Sing this two times:

Stop, stop, stop the waste
Of our H 2 O.
Use it only when you must
So it will keep its flow.

**Narrator**: Wally waves goodbye as the Water Savers fly off in the blue watercopter.

**Wally**: Well, I think I'll finish washing the car now. No sense in telling anyone about the Water Savers because they'll never believe me.     THE END

# ACTIVITIES FOR "WHAT DID YOU EAT FOR BREAKFAST?"

## Prereading Activities

Students can enter their vocabulary into their Science Picture Dictionaries or the teacher can orally review the words with the class before reading the play.

*carbohydrates   experimenting   fats   food pyramid*

*ingredients   minerals   nutrients   proteins   vitamins*

Ask students to illustrate and label what they ate for breakfast today. Have them list the food groups that are represented in their pictures.

## Purpose for Reading

Think about what you ate for breakfast today. Was it a healthy meal? Read *"What Did You Eat For Breakfast?"* to compare your breakfast to Alice, Willie, and Sue's breakfasts.

## Post-reading Activities

Use the Compare/contrast Activity Sheet to compare your breakfast to either Alice, Sue or Willie's breakfast. List at least three similarities and three differences. Write a sentence explaining how healthy your breakfast was. Compare and contrast Alice's and Willie's breakfasts on Sheet #2.

Make a class recipe book of healthy snacks. Your job is to make a healthy snack for your classmates. Write down the ingredients on the *My Super Snack Recipe Activity Sheet*. List the directions in order of how you made your tasty treat. Tell the class why your snack is healthy.

The class can sing the healthy food song at the end of the readers' theatre.

Primary Science Readers' Theatre

# ACTIVITY SHEET #1
# COMPARE/CONTRAST

Name _____

Breakfasts Being Compared:

_____

How are they different?

1. _____

2. _____

3. _____

4. _____

How are they the same?

1. _____

2. _____

3. _____

4. _____

© 2002 Pieces of Learning

Primary Science Readers' Theatre

# ACTIVITY SHEET #2

### Alice's Breakfast        Willie's Breakfast

How are they different?

1. _____

2. _____

3. _____

4. _____

How are they the same?

1. _____

2. _____

3. _____

4. _____

© 2002 Pieces of Learning

# ACTIVITY SHEET #3
# MY SUPER SNACK RECIPE

Student Name _____

Name of snack_____

Ingredients:

_____  _____

_____  _____

_____  _____

_____  _____

How to make my super snack:

1. _____

_____

2. _____

_____

3. _____

_____

4. _____

_____

My snack is healthy because _____

_____

# "WHAT DID YOU EAT FOR BREAKFAST?"
A play to teach about healthy eating and nutrition

Characters

Miss Nutrition    Alice    Narrator    Willie    Class    Sue

Vocabulary  carbohydrates, experimenting, fats, food pyramid, ingredients, minerals, nutrients, proteins, vitamins

SCENE 1    Miss Nutrition's class, Healthy Mountain Elementary School

**Narrator**: It's 10 a.m. and Miss Nutrition is teaching a lesson about healthy eating to her class.

**Miss Nutrition**: Sit down, Willie. That's the third time today you've been out of your seat, and it's only ten o'clock. Class, what are the six types of nutrients found in food?

**Class**: Carbohydrates, fats, proteins, water, vitamins, and minerals.

**Miss Nutrition**: Good, class! Sue, I noticed you weren't paying attention. You look very sleepy. When did you go to bed last night?

**Sue**: I always go to bed at nine o'clock. I don't know why I feel so sleepy.

**Miss Nutrition**: Willie, sit still please.

**Willie**: I'm trying to sit still, Miss Nutrition, but I just can't.

**Alice**: You're often jumpy in the mornings, Willie.

**Class**: Sue is mostly sleepy in the mornings.

**Miss Nutrition**: All right, class, what were we talking about?

**Class**: Nutrients.

**Miss Nutrition**: Oh, right! Now I remember, the six nutrients found in food. Why do we need all six nutrients, class?

**Narrator**: The class discusses the jobs of each nutrient.

**Class**: Carbohydrates and fats give us energy. Also, fats help our bodies use vitamins and fats keep us warm.

**Willie**: I love sweets! Are they carbohydrates? I hope so because I have lots of energy.

**Miss Nutrition**: I see that, Willie. Yes, sugar is a carbohydrate, but so are bread, cereal, rice, fruits and vegetables. Too much sugar isn't healthy for your body.

**Alice**: What about proteins? Aren't they good for our bodies?

**Miss Nutrition**: Yes, proteins help our bodies grow and work properly. Sue, wake up!

**Narrator**: Sue has her head on the desk and is fast asleep.

**Sue**: Huh? What? Where am I?

**Willie**: I'm thirsty. Can I get a drink of water?

**Miss Nutrition**: We'll take a break in ten minutes. When we eat our snacks, you can get a drink of water. Class, water is so important for healthy bodies. How much of our bodies is made of water?

**Class**: One third?

**Miss Nutrition**: No, over half of our bodies are made of water. We need water and food to survive.

**Alice**: What about vitamins and minerals? I take a chewy cherry dinosaur vitamin every morning with my breakfast.

**Miss Nutrition**: Vitamins and minerals are very important. Class, why do we need vitamins?

**Class**: Vitamins help our blood clot, and they help us use energy from foods we eat.

**Miss Nutrition**: Yes, and what about minerals?

**Class**: Minerals give us strong bones and teeth.

**Miss Nutrition**: Looks like you really know your nutrients, class. It's time for a five-minute break. You may eat your snacks and get drinks of water now.

**Narrator**: The class eats their snacks while Miss Nutrition talks to Sue.

SCENE 2      10:30 a.m., Miss Nutrition's classroom

**Narrator**: Sue seems to be more awake.

**Miss Nutrition**: Class, I think we can use the information about the six nutrients to help us eat healthier. Remember the food pyramid we studied last week?

**Class**: Yes, Miss Nutrition.

**Willie**: It has carbohydrates at the bottom. We can eat a lot of them. Yippee!! I love sugar, and you said it was a carbohydrate.

**Alice**: Willie, don't you remember that fats, sweets and oils are at the top of the food pyramid? The top is very small.

**Class**: We're not supposed to eat too many fats, sweets and oils.

**Willie**: Well, that's no fun!

**Miss Nutrition**: Willie, there are lots of delicious foods in the bread and cereal group. You can also have lots of fruit and vegetables.

**Willie**: I hate spinach!

**Sue**: What about Chocolate Puffies? I eat them for breakfast every morning.

**Alice**: I had oatmeal for breakfast today.

**Class**: Willie, what did you have for breakfast this morning?

**Willie**: Let's see. I ate a frosted blueberry breakfast tart. You know, the kind you put in the toaster.

**Miss Nutrition**: Well, class, I will make copies of the Nutrition Facts Labels from the breakfast tarts, Chocolate Puffies and the oatmeal. For homework, compare them, and see which ones are healthy breakfasts.

**Narrator**: The class puts the Nutrition Facts Labels into their homework folders.

SCENE 3    *The next day, 8:45 a.m., Miss Nutrition's class*

**Narrator**: The class is hanging up their coats, and everyone is talking about the homework assignment.

**Class**: What do you think about the Chocolate Puffies? How about the oatmeal? Are the breakfast tarts good for breakfast?

**Miss Nutrition**: I see that everyone is interested in last night's homework assignment. Let's discuss it.

**Willie**: Well, I wasn't surprised that the breakfast tart had 14 grams of sugar. It also had some vitamins and minerals. The 3 grams of protein were good, but I guess the 6 grams of fat were way too high. So, between all of that sugar and fat, no wonder I can't sit still. I have too much energy!

**Sue**: Well, my Chocolate Puffies had low fat, only one gram. That was way better than Willie's tart or Alice's oatmeal. They also have lots of vitamins and minerals. The Chocolate Puffies also had 4 grams of protein, which is good for growing bodies. However, the sugars were 19 grams, even worse than the breakfast tarts. The total carbohydrates was 43 grams!

**Miss Nutrition**: Sue, sometimes too many carbohydrates can make you sleepy.

**Sue**: I'm experimenting today. I ate an egg and whole wheat toast for breakfast. That's a lot more protein and less sugar and carbohydrates. Let's see if I get sleepy today.

**Class**: It looks like oatmeal is the winner.

**Alice**: What a surprise! The unsweetened oatmeal had 3 grams of fat, 26 grams of carbohydrates and 5 grams of protein. It's a good thing that I like to eat oatmeal.

**Miss Nutrition**: Now that we've learned how to read Nutrition Facts Labels, we will all make healthy snacks. Your homework is to bring in a recipe of a healthy snack. Remember to use the food pyramid and nutrition labels to help you. When it is your turn, you will bring in enough of your snack for all of us to try. Write down your recipe and its nutrition facts. We'll decide if it's a healthy snack. Each day will be another person's turn.

**Class**: Can we make a recipe book of healthy snacks?

**Miss Nutrition**: Let's call our book <u>Super Snacks</u>!

**Willie**: Miss Nutrition, can we sing our healthy food song?

**Class**: Yes, let's sing about eating right and feeling good.

*(To the tune of "She'll Be Comin' Round the Mountain")*

We'll be using good nutrition when we eat, yahoo!
We'll be using good nutrition when we eat.
We'll be using good nutrition, we'll be using good nutrition,
We'll be using good nutrition when we eat.

We will read food labels at the store, yahoo!
We will read food labels at the store.
We will read food labels, we will read food labels,
We will read food labels at the store.

We will use the food pyramid everyday, yahoo!
We will use the food pyramid every day.
We will use the food pyramid, we will use the food pyramid,
We will use the food pyramid everyday.

We will drink lots of water, yes we will, yahoo!
We will drink lots of water, yes we will.
We will drink lots of water, we will drink lots of water,
We will drink lots of water, yes we will.

We'll be using good nutrition when we eat, yahoo!
We'll be using good nutrition when we eat.
We'll be using good nutrition, we'll be using good nutrition,
We'll be using good nutrition when we eat.

**Miss Nutrition**: It looks like you are on your way to healthy eating.

**Class**: Did you notice that Sue is still awake?

**Sue**: I feel very alert, thanks to all that I've learned in school. I'll miss my Chocolate Puffies, but I don't feel sleepy anymore.

**Miss Nutrition**: Class, have fun planning your healthy snacks.

### THE END

# ACTIVITIES FOR "WHY MATTER MATTERS"

## Prereading Activities

Students can enter their vocabulary into their Science Picture Dictionaries or the teacher can orally review the words with the class before reading the play.

condense    evaporate    heat    matter

texture    water vapor    freeze    melt

Draw what you ate for breakfast this morning. Put an "s" next to the solids, an "l" next to the liquids, and a "g" next to the gases.

## Purpose For Reading

To learn more about solids, liquids and gases. You will find out if you labeled your breakfast correctly.

## Post-reading Activities

Go back to your breakfast drawing and check the labels. Change any that are not correct. On the back of your drawing, explain how you know that what you ate are solids, liquids or gases. What is your proof?

What questions would you ask Miss Matter? Write at least three questions about matter. When you have finished, your teacher will divide your class into two teams. She will use everyone's questions to ask the teams. Which team will be the Matter Champions?

Use the Activity Sheet for *Why Matter Matters* to show how water can exist in all three stages of matter.

Primary Science Readers' Theatre

# ACTIVITY SHEET
# "WHY MATTER MATTERS"

Name _____

Draw water as a solid, a liquid and a gas.

|  |  |
|---|---|
| water as a solid | water as a liquid |

water as a gas

First when the water was a solid it was called _____.

Then it changed to a liquid because _____

_____.

After that, it changed into a gas because _____

_____.

So, the reason water can change to another form of matter is

because _____

_____.

© 2002 Pieces of Learning

Primary Science Readers' Theatre

# "WHY MATTER MATTERS"
A play to teach the forms of matter

Characters
Miss Matter, the teacher     Narrator     Team 1     Team 2 (the class)

Vocabulary     condense, evaporate, freeze, heat, matter, melt, texture, water vapor

*Scene One          Vaportown Elementary School*

**Narrator**: Miss Matter is teaching a science lesson about solids, liquids, and gases.

**Teacher**: Class, as you know, we have been studying about matter. Matter is anything that takes up space, such as solids, liquids, and gases. To practice for our quiz, I'm going to ask you some questions. Just answer true or false. Ready?

**Class**: Yes, Miss Matter.

**Teacher**: Question number 1: Solid matter does NOT change shape or size.

**Class**: True.

**Teacher**: Good! Question number 2: Liquids, gases and solids all have texture.

**Class**: True- er- false- huh?

105                    © 2002 Pieces of Learning

**Teacher**: I see you're not too sure about that one. Texture is the way something feels. Gas doesn't have a texture, but solids and liquids have texture. Question number 3: Liquids change shape depending on the shape of their container.

**Class**: True.

**Teacher**: Very good! Question number 4: All liquids feel alike.

**Class**: False.

**Teacher**: Right! Milk and oil have very different textures. Question number 5: Air is a gas that is all around us.

**Class**: False-er-true-huh?

**Teacher**: It is TRUE that air is a gas and that it IS all around us. We just can't see it. Here's another one. Question number 6: Air takes up space.

**Class**: True.

**Teacher**: I'm glad you remember how air fills balloons and plastic bags. I'm proud of you. Just one more question: Matter can change from liquid to gas or from gas to liquid.

**Class**: Er-false-true-huh?

**Teacher**: Last weekend the water in the glass evaporated. That means the water became a gas or became water vapor. That is liquid turning into gas. Remember?

**Class**: Yes, Miss Matter.

**Teacher**: Rain is just water vapor being cooled and changing into liquid. That's called condensing. Bet you can't say condensing three times fast.

**Class**: Condensing, condensing, condensing!

**Teacher**: Let's divide into two teams. We'll go outside on a Matter Hunt. Each team needs to find one solid, one liquid, and one gas. Whichever team does it first correctly will be the winner. Bring containers with you. All set?

**Class**: You bet!

**Narrator**: The class follows Miss Matter outside to begin the Matter Hunt.

*Scene Two*
*On the playground*

**Team 1**: Let's go behind the school away from the swings so we can keep our matter a secret.

**Team 2**: We'll go over by the jungle gym. Let's talk quietly so team one won't hear us.

**Narrator**: Each team begins exploring.

**Team 1**: Here's a branch that fell from the maple tree. It's a solid.

**Team 2**: There's a melted crayon on the blacktop. Now that it cooled off, it's a solid again.

**Narrator**: It begins to rain.

**Teacher**: Hurry, class! The water vapor is condensing on our heads!

**Team 2**: Quick! Put the bowl on the ground to catch the rain water. Let's go hunt for a gas.

**Team 1**: Here's some water in a puddle. We'll collect it in this container. Hurry and look for a gas.

**Narrator**: Just then Team 2 remembered that Billy's bicycle tires had air in them, so he ran to get his bicycle.

**Team 2**: Miss Matter, we have air in the bicycle tires, a crayon and rainwater. That's a gas, a solid and a liquid.

**Teacher**: Team 1, what do you have?

**Team 1**: We have a tree branch for a solid and water for a liquid. We didn't find a gas.

**Teacher**: Air is a gas that is all around us. Well, it looks like Team 2 wins. Let's go back inside. We're getting wet.

*Scene Three*
*Back in the classroom*

**Narrator**: The wet students return, carrying all their matter.

**Teacher**: We can have a special treat tomorrow. Team 2 can pick their favorite solids, liquids and gases.

**Team 2**: We want apple juice for the liquid. We'll eat soft pretzels for the solids. We'd like ice cream, too. That's a frozen liquid, but now it's a solid. We'll decorate with balloons for the gas.

**Narrator**: Each team made decorations from solid matter.

**Team 1**: This is fun, even if our team didn't win.

**Teacher**: Class, I have a special song for you about matter.

**Class**: Hooray!

(*To the tune of "You Are My Sunshine"*)

You are my solid, you never change shape,
You have texture, color and size.
You'll stay a solid unless you're heated,
So please stay a solid, that's wise.

    You are a liquid, and you have texture,
    You can be solid if you freeze.
    You come in colors and your shape changes,
    So stay a liquid, won't you please?

        You are a gas and we can't see you,
        You change to liquids when you cool.
        You can fill bubbles, balls and balloons,
        So please stay a gas, THAT'S COOL!!

    **Teacher**: Well, class, now you know why matter matters!

**THE END**

# Notes, Resources, & Information

Alfie Astronaut Visits the Moon

The Case of the Spreading Germs

Dinosoccer

How Flossy Flamingo Lost Her Habitat

How Rachel Rabbit Fixed the Food Chain

The Particles Visit Energyland

Stop! You're Interrupting My Life Cycle

The Sunbeams Take a Trip

## Wally the Water Waster

## What Did You Eat For Breakfast?

## Why Matter Matters